ROTARY DRILLING SERIES

Drilling a Straight Hole

Unit II, Lesson 3
Third Edition

By William E. Jackson

Published by

PETROLEUM EXTENSION SERVICE
Continuing & Extended Education
The University of Texas at Austin
Austin, Texas

in cooperation with

INTERNATIONAL ASSOCIATION
OF DRILLING CONTRACTORS
Houston, Texas

2000

Library of Congress Cataloging-in-Publication Data

Jackson, William E., 1929—
Drilling a straight hole / by William E. Jackson. — 3rd ed.
p. cm. — (Rotary drilling series ; unit 2, lesson 3)
ISBN 0-88698-193-X
1. Oil well drilling. 2. Oil well drilling rigs. I. Title. II. Series.
TN871.2.J317 2000
622'.3382—dc21 00-010331
CIP

First Edition published 1968. Second Edition 1982.
Third Edition 2000. Third Impression 2008
Printed in the United States of America

Catalog no. 2.20330
ISBN 0-88698-193-X

Contents

Figures

Tables

Foreword

For many years, the Rotary Drilling Series has oriented new personnel and further assisted experienced hands in the rotary drilling industry. As the industry changes, so must the manuals in this series reflect those changes.

The revisions to both the text and illustrations are extensive. In addition, the layout has been "modernized" to make the information easy to get; the study questions have been rewritten; and each major section has been summarized to provide a handy comprehension check for the reader.

PETEX wishes to thank industry reviewers—and our readers—for invaluable assistance in the revision of the Rotary Drilling Series.

While horizontal drilling has made great progress in recent years, the principles behind keeping a hole straight and on course still apply to the work of a rig crew. Bottomhole assemblies have never been more important in drilling a successful hole as they are now; this book discusses them at length. MWD techniques have made maintaining a hole on course easier than ever and the book explains the principle as well. In short, this new edition of *Drilling a Straight Hole* should be a welcome addition to anyone's library.

Although every effort was made to ensure accuracy, this manual is intended to be only a training aid; thus, nothing in it should be construed as approval or disapproval of any specific product or practice.

Ron Baker

Acknowledgments

The author expresses a sincere appreciation to the many people who contributed to this edition of *Drilling a Straight Hole.* Those who provided illustrations, background information or discussion on the subject include:

John Baer, Division Engineer, Helmerich & Payne International Drilling Company, Oklahoma City, OK.

R. L. Hilbun, Licensed Professional Engineer and owner, Summa Engineering Inc., Oklahoma City, OK.

Robert Gum, Applications Design Engineer, Security DBS, Oklahoma City Division, Oklahoma City, OK.

Jeff Hubbard, District Manager, Drilco Group, Smith Services, Oklahoma City District, Oklahoma City, reviewed the manuscript and also provided many illustrations and useful background material.

Chuck Henkes, Account Representative with Sperry-Sun Drilling Services in Oklahoma City, was especially generous with his time and pertinent material. His critical manuscript reviews and suggestions contributed greatly to the final product.

Others who provided illustrations and permission to publish their material include:

- Smith Bits Division of Smith International, Inc.
- Baker Hughes INTEQ
- Drilco Grant Drilling Handbook, Smith International, Inc.
- IADC
- Society of Professional Engineers (SPE)

The staff at the Oklahoma Commission on Marginal Wells, Norman, Oklahoma, provided generous use of material from their excellent library.

A sincere thanks is extended to all who helped.

Units of Measurement

Throughout the world, two systems of measurement dominate: the English system and the metric system. Today, the United States is almost the only country that employs the English system.

The English system uses the pound as the unit of weight, the foot as the unit of length, and the gallon as the unit of capacity. In the English system, for example, 1 foot equals 12 inches, 1 yard equals 36 inches, and 1 mile equals 5,280 feet or 1,760 yards.

The metric system uses the gram as the unit of weight, the metre as the unit of length, and the litre as the unit of capacity. In the metric system, for example, 1 metre equals 10 decimetres, 100 centimetres, or 1,000 millimetres. A kilometre equals 1,000 metres. The metric system, unlike the English system, uses a base of 10; thus, it is easy to convert from one unit to another. To convert from one unit to another in the English system, you must memorize or look up the values.

In the late 1970s, the Eleventh General Conference on Weights and Measures described and adopted the Système International (SI) d'Unités. Conference participants based the SI system on the metric system and designed it as an international standard of measurement.

The *Rotary Drilling Series* gives both English and SI units. And because the SI system employs the British spelling of many of the terms, the book follows those spelling rules as well. The unit of length, for example, is *metre*, not *meter*. (Note, however, that the unit of weight is *gram*, not *gramme*.)

To aid U.S. readers in making and understanding the conversion to the SI system, we include the following table.

English-Units-to-SI-Units Conversion Factors

Quantity or Property	English Units	Multiply English Units By	To Obtain These SI Units
Length, depth, or height	inches (in.)	25.4	millimetres (mm)
		2.54	centimetres (cm)
	feet (ft)	0.3048	metres (m)
	yards (yd)	0.9144	metres (m)
	miles (mi)	1609.344	metres (m)
		1.61	kilometres (km)
Hole and pipe diameters, bit size	inches (in.)	25.4	millimetres (mm)
Drilling rate	feet per hour (ft/h)	0.3048	metres per hour (m/h)
Weight on bit	pounds (lb)	0.445	decanewtons (dN)
Nozzle size	32nds of an inch	0.8	millimetres (mm)
Volume	barrels (bbl)	0.159	cubic metres (m^3)
		159	litres (L)
	gallons per stroke (gal/stroke)	0.00379	cubic metres per stroke (m^3/stroke)
	ounces (oz)	29.57	millilitres (mL)
	cubic inches ($in.^3$)	16.387	cubic centimetres (cm^3)
	cubic feet (ft^3)	28.3169	litres (L)
		0.0283	cubic metres (m^3)
	quarts (qt)	0.9464	litres (L)
	gallons (gal)	3.7854	litres (L)
	gallons (gal)	0.00379	cubic metres (m^3)
	pounds per barrel (lb/bbl)	2.895	kilograms per cubic metre (kg/m^3)
	barrels per ton (bbl/tn)	0.175	cubic metres per tonne (m^3/t)
Pump output and flow rate	gallons per minute (gpm)	0.00379	cubic metres per minute (m^3/min)
	gallons per hour (gph)	0.00379	cubic metres per hour (m^3/h)
	barrels per stroke (bbl/stroke)	0.159	cubic metres per stroke (m^3/stroke)
	barrels per minute (bbl/min)	0.159	cubic metres per minute (m^3/min)
Pressure	pounds per square inch (psi)	6.895	kilopascals (kPa)
		0.006895	megapascals (MPa)
Temperature	degrees Fahrenheit (°F)	$\frac{°F - 32}{1.8}$	degrees Celsius (°C)
Thermal gradient	1°F per 60 feet	—	1°C per 33 metres
Mass (weight)	ounces (oz)	28.35	grams (g)
	pounds (lb)	453.59	grams (g)
		0.4536	kilograms (kg)
	tons (tn)	0.9072	tonnes (t)
	pounds per foot (lb/ft)	1.488	kilograms per metre (kg/m)
Mud weight	pounds per gallon (ppg)	119.82	kilograms per cubic metre (kg/m^3)
	pounds per cubic foot (lb/ft^3)	16.0	kilograms per cubic metre (kg/m^3)
Pressure gradient	pounds per square inch per foot (psi/ft)	22.621	kilopascals per metre (kPa/m)
Funnel viscosity	seconds per quart (s/qt)	1.057	seconds per litre (s/L)
Yield point	pounds per 100 square feet (lb/100 ft^2)	0.48	pascals (Pa)
Gel strength	pounds per 100 square feet (lb/100 ft^2)	0.48	pascals (Pa)
Filter cake thickness	32nds of an inch	0.8	millimetres (mm)
Power	horsepower (hp)	0.75	kilowatts (kW)
Area	square inches ($in.^2$)	6.45	square centimetres (cm^2)
	square feet (ft^2)	0.0929	square metres (m^2)
	square yards (yd^2)	0.8361	square metres (m^2)
	square miles (mi^2)	2.59	square kilometres (km^2)
	acre (ac)	0.40	hectare (ha)
Drilling line wear	ton-miles (tn•mi)	14.317	megajoules (MJ)
		1.459	tonne-kilometres (t•km)
Torque	foot-pounds (ft•lb)	1.3558	newton metres (N•m)

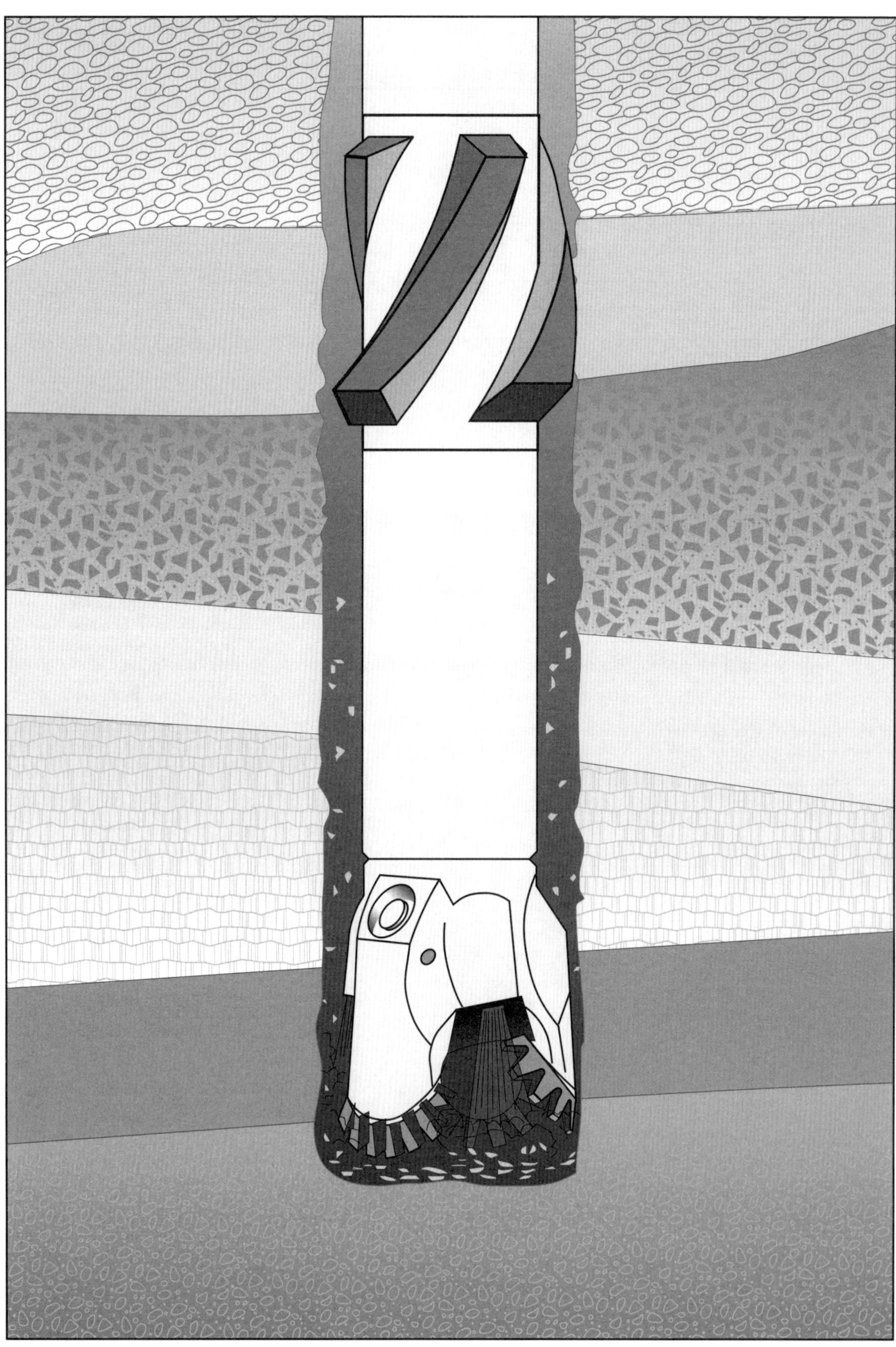

Introduction

When rotary drilling first began, operators and drillers assumed that if they held the kelly vertical when starting the hole, the drill string and bit would drill a straight hole. In the Seminole boom in Oklahoma in 1928-29 however, the industry began to suspect that holes were crooked. On occasion, wellbores actually intersected. In addition, actual drilled depths did not correspond to projected formation depths. Obviously, the Oklahoma rigs were drilling crooked holes.

Crooked holes were not just a problem for geologists. On the contrary, a crooked hole involved drilling more footage when compared to a straight hole (fig. 1); consequently the operator had to pay for the extra footage. Furthermore, contractors sometimes drilled into existing boreholes or producing wells on offset leases, creating serious legal problems. Hole deviation thus became such an important consideration that operators began looking for ways to determine the amount of downhole deviation needed.

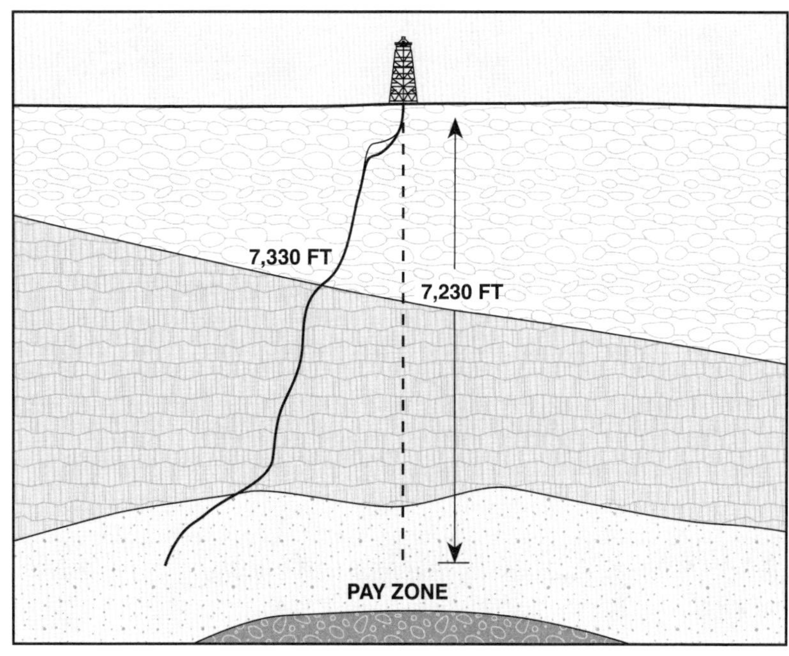

Figure 1. Drilled depth in a crooked hole versus a true vertical depth. More hole must be drilled in a crooked hole than in a straight hole to reach the same formation.

Development of Wellbore Survey Tools

In 1929, the first deviation-recording tool was developed (fig. 2) and later, many drilling contracts included a deviation clause that required the contractor to maintain the hole angle within a specified maximum limit. Early contracts were quite liberal; 10 to 15 degrees from vertical was acceptable. These deviation restrictions did not reduce costs as much as operators expected because they did not reduce the extra time or footage drilled. Thus, by 1935 most contract deviation clauses called for 2- or 3-degree maximum hole deviation.

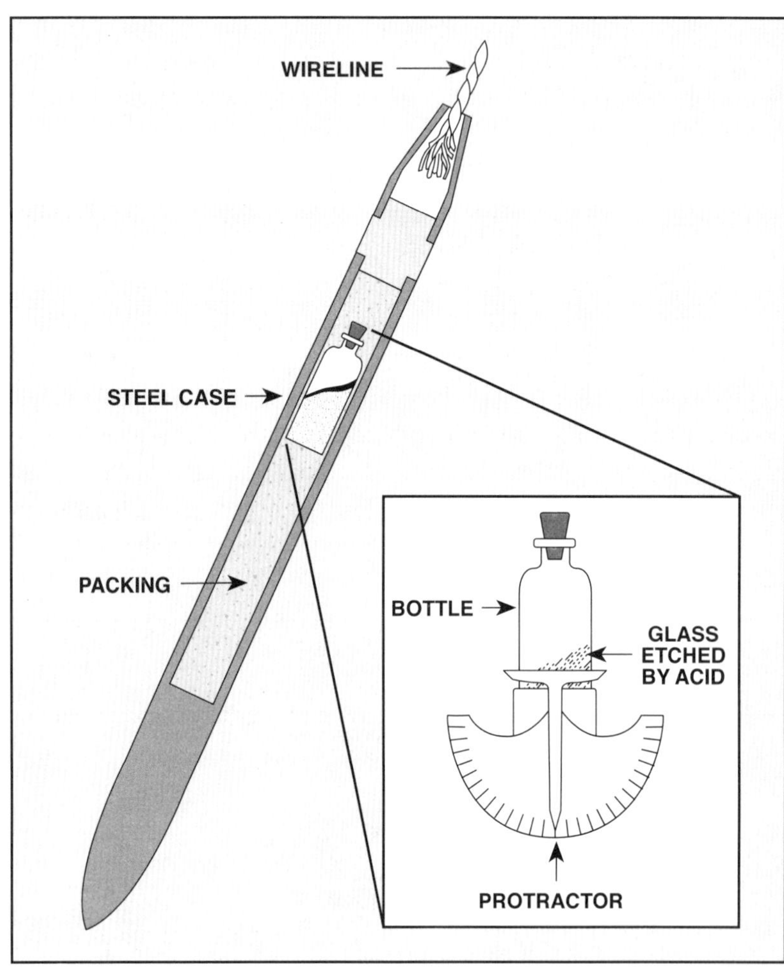

Figure 2. An early borehole surveying instrument. The acid bottle was one of the earliest surveying instruments. A glass bottle was filled with about ½ oz. (15 millilitres) of diluted hydrofluoric acid topped with a layer of oil. It was placed in a housing, lowered down the drill pipe to the bit, and allowed to stand until the acid etched a line inside the bottle. Upon retrieval, the angle of inclination indicated by the etched line on the bottle was read with the use of a protractor.

Early well-surveying instruments, such as the acid or ink bottle designs (fig. 3), could measure the amount of deviation but not the direction of deviation.

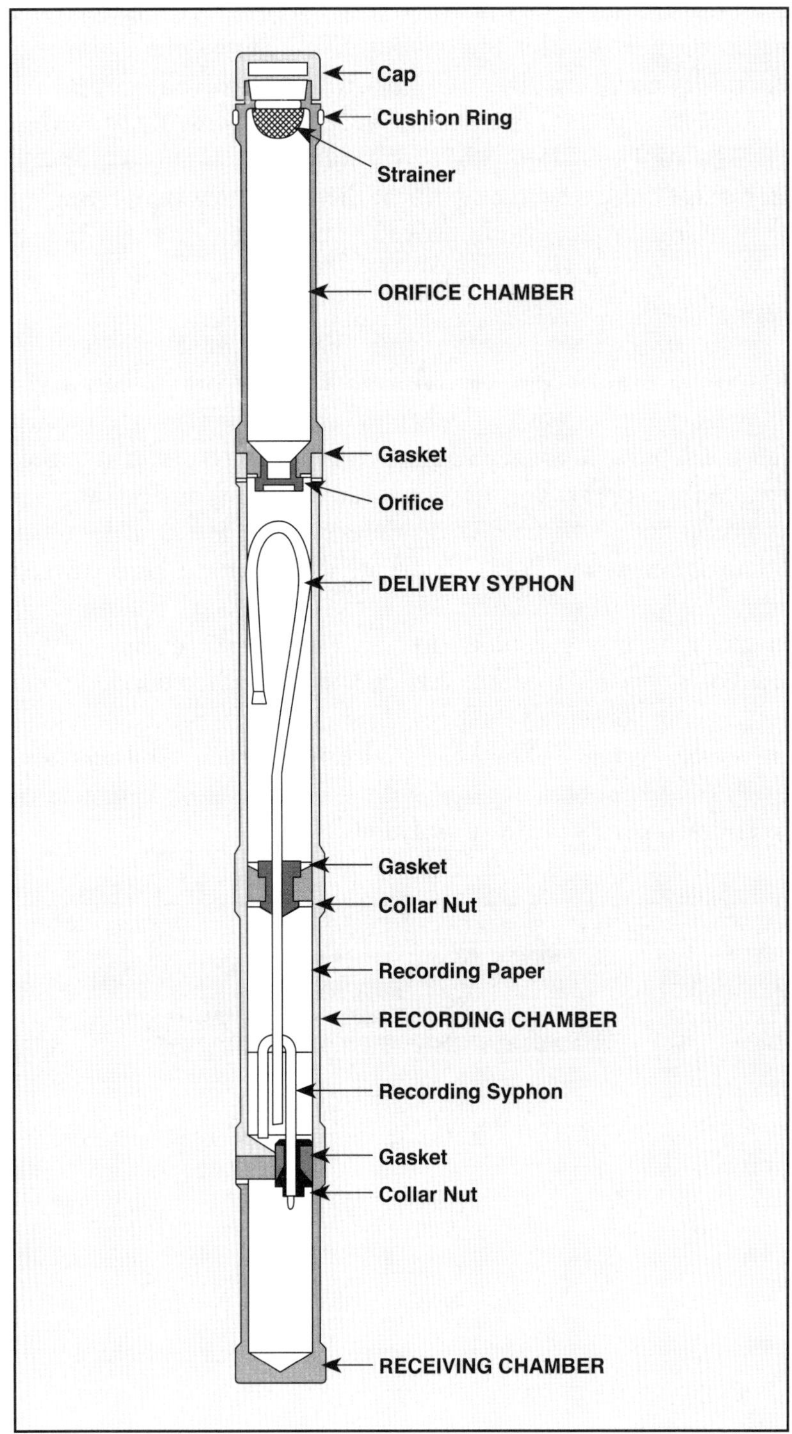

Figure 3. An ink-bottle survey instrument called the SYFO inclinometer consists of four chambers. The orifice chamber is filled with ink. As the instrument within its protective case is lowered into the hole, the ink flows into the delivery chamber at a regulated rate. The ink is then siphoned into the recording chamber where it will leave a wavy line on graph paper. The ink is then stored in the receiving chamber until the instrument is retrieved. (Courtesy of Sperry-Sun Drilling Services, a Halliburton Company)

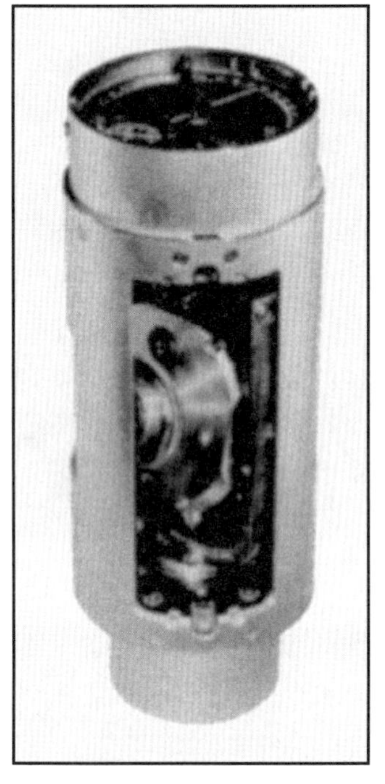

Figure 4. Gyroscope no. 1 designed and built by Elmer Sperry, Jr. in 1929. This gyro was in service for 36 years and surveyed more than 4 million feet (6 million kilometres) of hole. (Courtesy of Sperry-Sun Drilling Services, a Halliburton Company)

Also in 1929, Sperry Gyroscope Company and Sun Oil Company developed the first survey tool using a gyroscope and camera (fig. 4). The gyroscope maintained a north orientation and, in combination with an inclinometer and a camera, recorded both direction and inclination of the borehole. Given the trade name Surwell by the Sperry-Sun Well Surveying Company, operators could run the tool on wireline, drop and recover it by overshot, or pull it from the hole with the drill string (fig. 5). The tool could quickly make multiple readings at regular intervals on the way in and repeat them on the way out. Survey results from many wells proved conclusively that wellbores twisted and turned in a spiral, and often made abrupt turns at formation boundaries.

In 1932, a well survey at the Huntington Beach field in California confirmed a lease violation. A company intentionally allowed its wells to drift as much as 1,500 feet (ft) or 460 metres (m) at a depth of 4,000 ft or 1,200 m (fig. 6). The ensuing litigation resulted in a court ruling that halted future drilling until authorities could survey all the field wells. Subsequently, a statewide ruling required that new wells in California had to conform to a restricted allowable inclination. This case was the first instance that required operators to take borehole surveys. It was legal recognition that wellbores deviated from the vertical and established borehole surveys as court-accepted evidence. Later, many cases used borehole surveys as evidence. The largest occurred in the East Texas field in 1962 where court-ordered surveys confirmed 90 instances of intentional slant-hole lease violations.

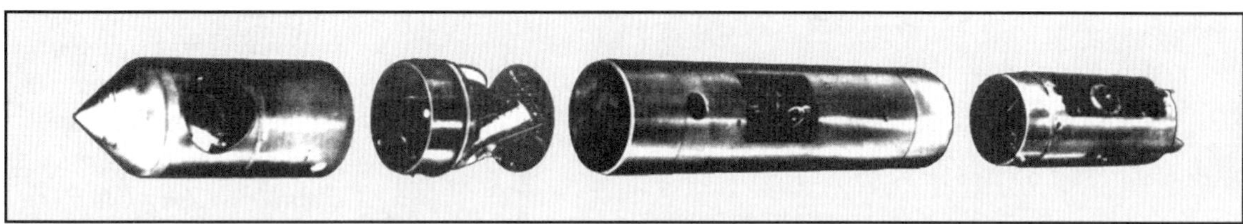

Figure 5. Components in a 5½" (139.7 mm) SURWELL well-surveying instrument (Courtesy of Sperry-Sun Drilling Services, a Halliburton Company)

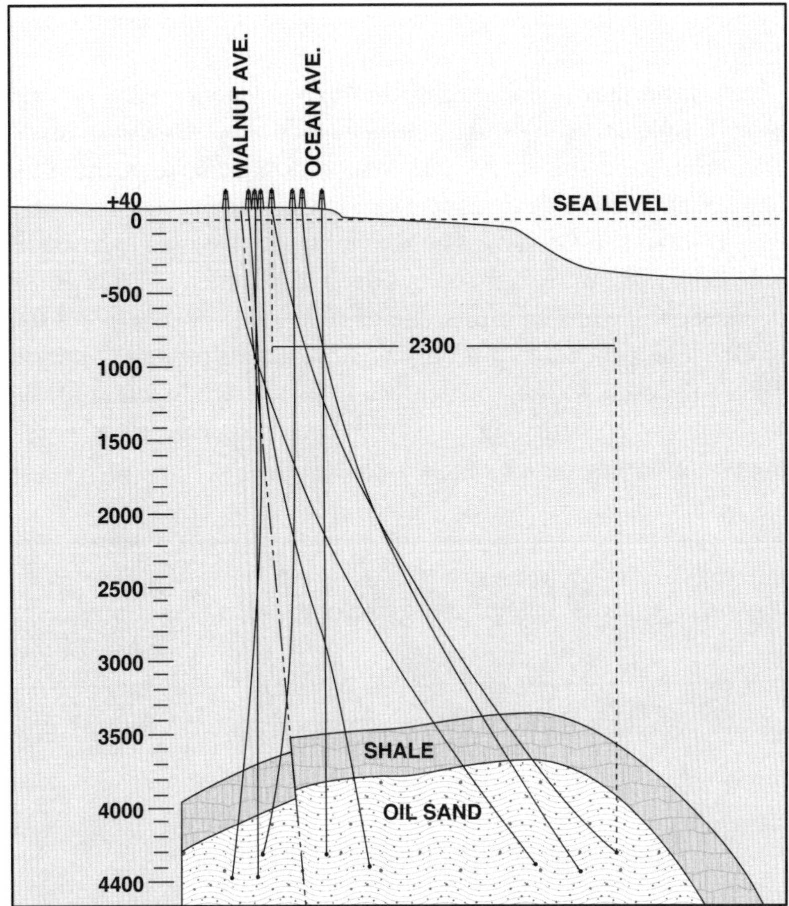

Figure 6. Deflected drilling at Huntington Beach, California, 1935 (Courtesy of Sperry-Sun Drilling Services, a Halliburton Company)

Through the years, borehole survey instruments have undergone many refinements, such as their speed of operation, reduction in the size of the gyroscopes, accuracy of readings, design of the magnetic instrumentation, and increases in capability. Today, sophisticated electronic instruments can monitor borehole direction and inclination while drilling proceeds. Operators commonly use steerable systems consisting of a downhole mud motor and measurement-while-drilling (MWD) telemetry with surface monitors.

Contract Deviation Clauses

By 1935, with the validity of borehole surveys firmly established, drilling contracts included deviation clauses allowing 2 to 3 degrees as the maximum permissible hole deviation. After World War II, an upsurge in drilling the deep, hard-rock basins of Texas, Oklahoma, and the Rocky Mountains occurred. To maintain a good penetration rate in these areas, drillers simply applied more weight on the bit (WOB). The result, however, was an increase in hole deviation, and when surveys determined the hole was going to exceed the contract limits, the usual solution was to lower the WOB. This procedure not only slowed the drilling rate, but also often produced a *dogleg*, which is a rapid change in angle of the wellbore (fig. 7). A dogleg can cause costly problems; at its worst, the operator may have to redrill the hole.

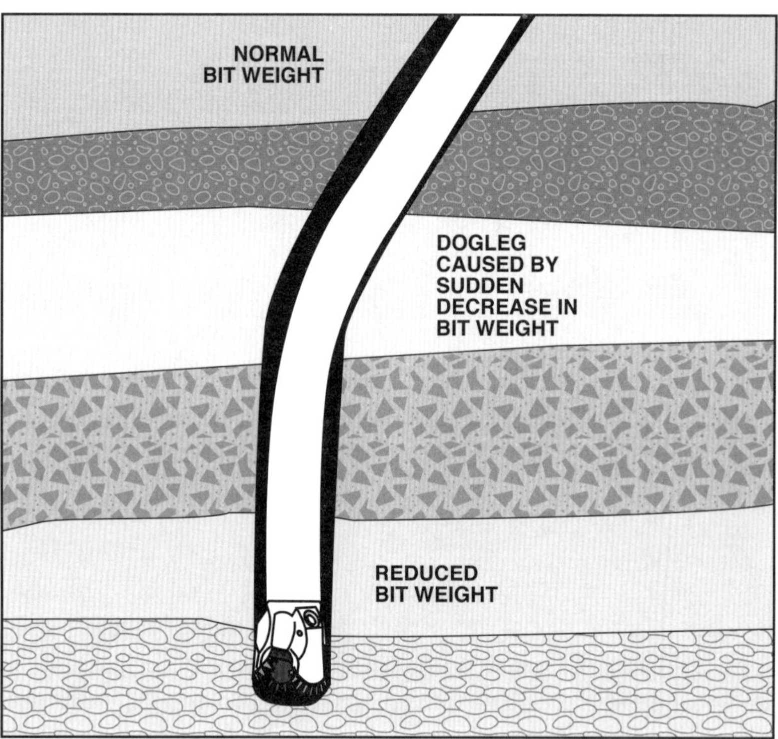

Figure 7. A dogleg caused by reduced WOB. Reduced weight can be used to reduce hole angle but a sudden reduction often produces a dogleg.

By 1950, operators and contractors could see that strict adherence to the 3-degree limitation in deviation clauses was not always beneficial. Slow penetration rates and plug-back operations to straighten the hole often made drilling uneconomical. Drilling engineers maintained that operators could economically justify many more wells if they allowed more latitude in deviation clauses. They argued that the total deviation was not as important as controlling the deviation in a given number of feet (metres) of hole. For instance, a 6-degree deviation occurring in 500 ft (150 m) of hole would not cause a problem. However, a change of 3 degrees occurring in 100 ft (30 m) of hole could produce a dogleg that could render the hole unusable.

Modern drilling contracts include more liberal deviation clauses that permit a usable hole to be drilled at the lowest cost. Today, operators include several provisions in deviation clauses, such as permitting greater hole angle near the bottom of the hole or below certain depths or formations (fig. 8). A contract may call for the use of air percussion tools or air drilling methods that can assure a straighter, lower-cost hole in problem areas.

3. STRAIGHT HOLE SPECIFICATIONS (See Subparagraph 9.4)

Well Depth		Maximum Distance Between Surveys,	Maximum Deviation from Vertical	Maximum Change of Inclination per 100',
From	To	Feet	Degrees	Degrees (1)
______	______	______	______	______
______	______	______	______	______
______	______	______	______	______
______	______	______	______	______
______	______	______	______	______
______	______	______	______	______

Location of wellbore at ________ feet shall be ____

____

____

____

Figure 8. The deviation clause from a drilling contract (Courtesy of IADC)

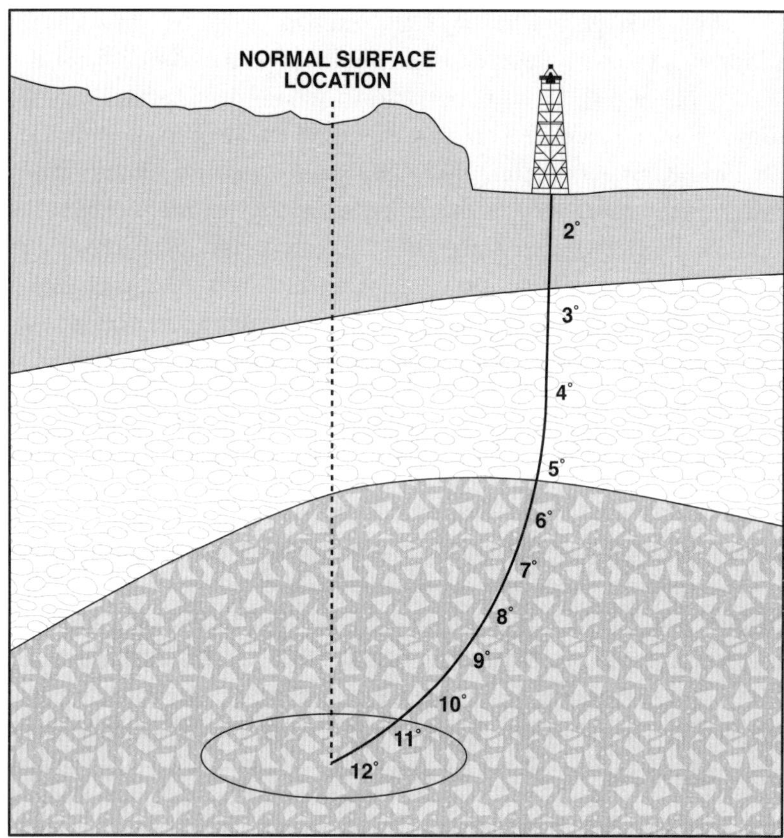

Figure 9. A surface location selected to allow the bit to drift into the target area permits faster penetration to total depth. Note the increased deviation near the bottom of the hole.

In addition, the operator can select a surface location that allows the bit to drift naturally (updip) into the target area (fig. 9).

New technology allows much greater control of hole deviation. Today, deviation can be routinely monitored and steered in a deviated hole while drilling it. Downhole drilling motors with MWD capability, using improved bit and drill string designs, can drill straight, usable holes faster and more efficiently than ever.

To summarize—

Crooked holes

- mean drilling more footage than necessary to reach the objective
- can cause serious legal problems

Wellbore survey tools

- first developed in 1929
- earliest designs were acid or ink-bottle types and measured only deviation, not direction
- Sperry Gyroscope and Sun Oil developed gyroscopic tools to measure both deviation and direction
- sophisticated electronic instruments and MWD tools often used today

Contract deviation clauses

- at first, restricted deviation to 2 to 3 degrees
- later, allowed for more liberal deviation, especially near the bottom of the hole.

Straight Hole Considerations

The term straight hole loosely describes a borehole that a drilling contractor has drilled vertically, from top to bottom. In reality, practically all wellbores deviate from the vertical. It is virtually impossible to drill a perfectly straight hole. Drilling contracts recognize this fact and allow a variation from the strict term. A better description of modern drilling practices is controlled deviation drilling because industry now accepts a straight hole as one that meets two qualifications:

1. The hole stays within the boundary of a cone, as designated by the operator in the deviation clause of the contract. The total hole angle is therefore restricted (fig. 10).
2. The hole does not change direction rapidly, usually no more than 3 degrees per 100 ft (30 m) of hole. The rate of hole-angle change is therefore restricted.

Staying within these allowable parameters, the contractor's main objective is to deliver a straight and usable hole to the specified depth. The usable borehole should be full gauge, smooth, free of doglegs, keyseats, ledges, offsets, and spirals that permit completion and production operations that are free from trouble.

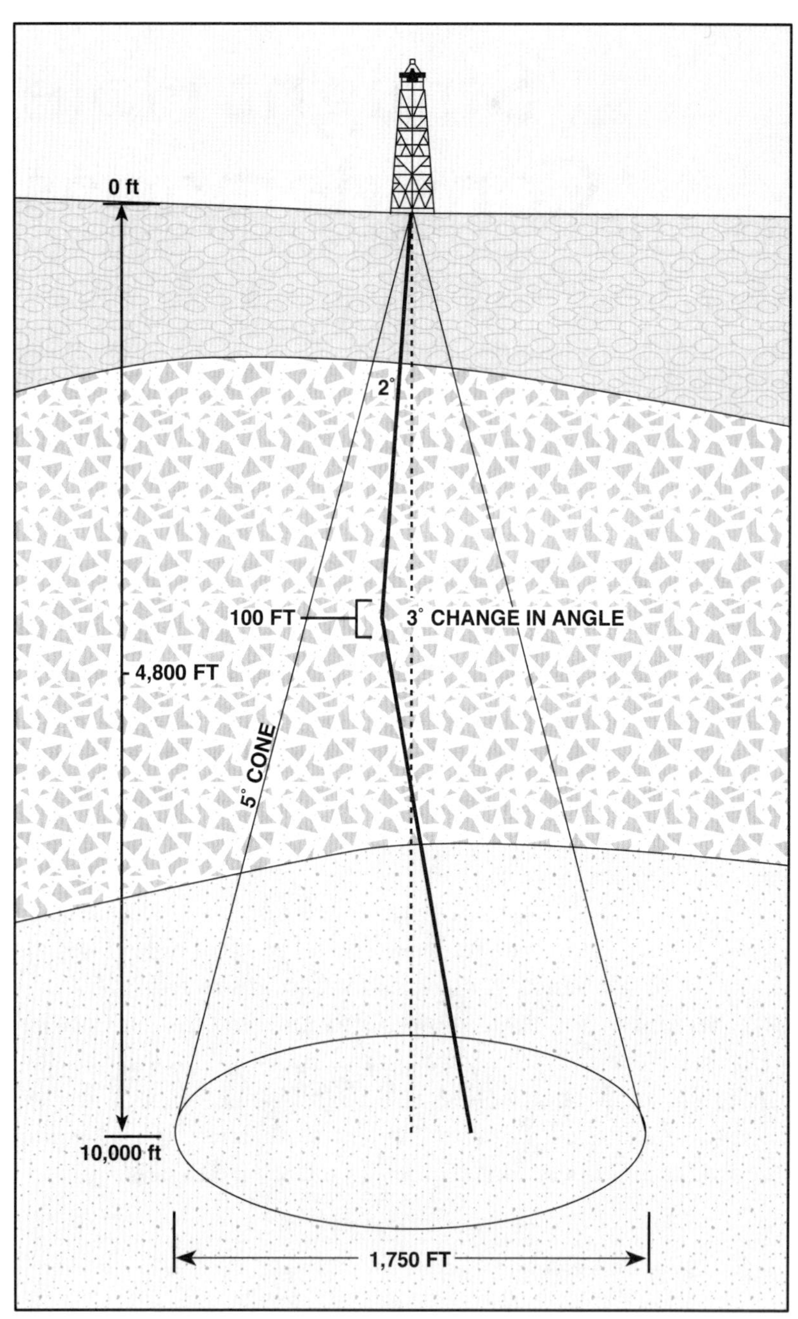

Figure 10. A straight hole. Specifications are for a 10,000-foot (3,048-metre) well with a 5-degree maximum deviation, making the target area a 1,750-ft (533-metre) circle directly beneath the rig floor. A 3-degree specification would reduce the target area to a diameter of 1,048 ft (319 metres). Only one change in angle is shown but an actual well may have many changes in angle and still be accepted as a straight hole.

Restricted Total Hole Angle

Reasons to restrict total hole angle (fig. 11) include:

1. ensuring that the hole will bottom out within a reasonable horizontal distance from the surface location;
2. keeping costs down by not having to drill extra footage when the hole drifts outside the restricted hole angle;
3. keeping the wellbore within the lease boundary so that no legal questions arise as to ownership of the hydrocarbons produced from the well;
4. keeping the wellbore within the distances from lease or unit lines established;

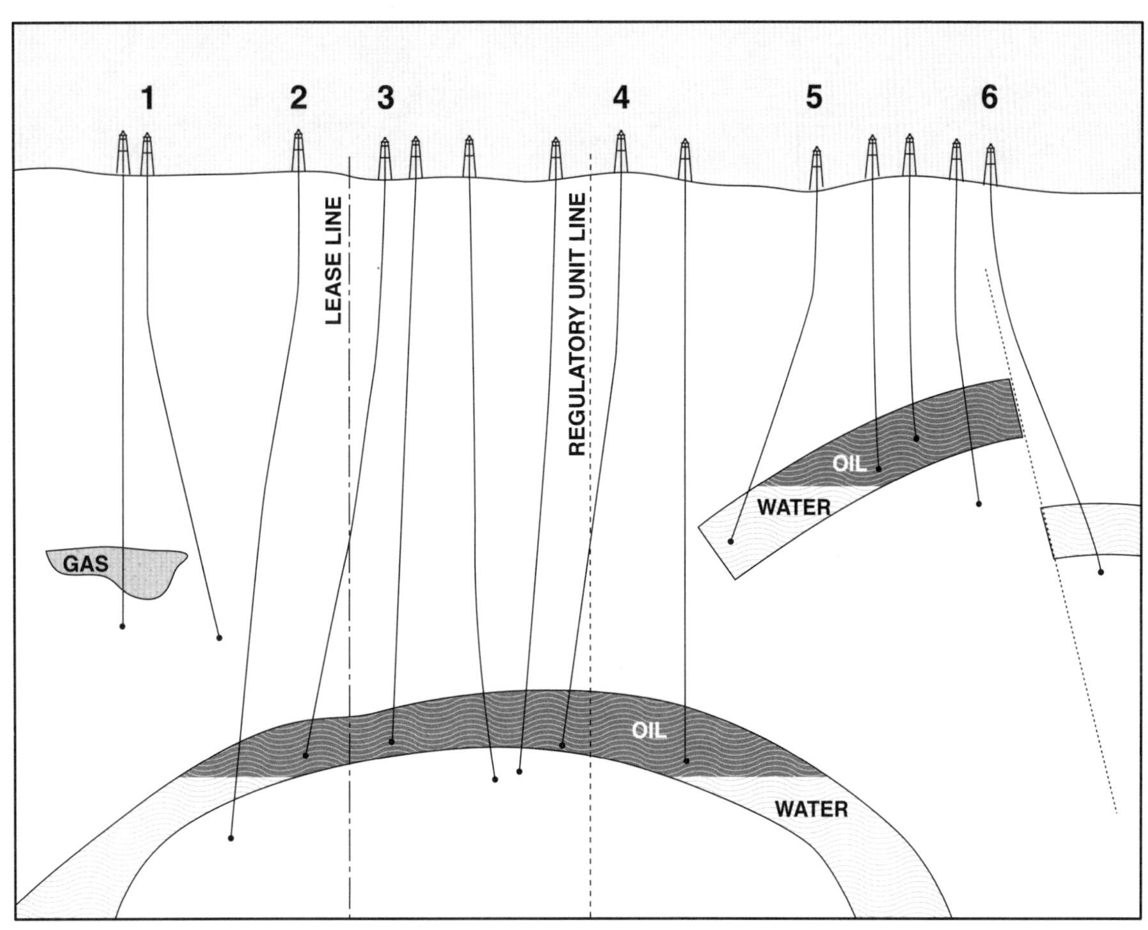

Figure 11. Reasons to restrict total hole angle: 1. dry hole missed narrow gas sand and drilled extra hole; 2. oilwell will probably go to water early, producing less oil; 3. productive well violated lease boundary, produces oil illegally; 4. productive well violated regulatory unit line, may be shut in; 5. dry hole wellbore drifted into water-filled part of reservoir; 6. dry hole wellbore drifted to wrong side of fault, missed pay, drilled extra hole.

5. assuring that the hole penetrates a specific pay zone (the target zone may be restricted to a small fault block or a narrow sand lens in a stratigraphic trap); and
6. assuring that the hole does not drift to penetrate the objective formation at a point below the oil-gas-water contact, resulting in a dry hole.

Restricting the total hole angle does not automatically ensure a straight hole. The typical 5-degree limitation does not guarantee that a wellbore will be free of doglegs that may even exceed the 5 degrees allowed (fig. 12).

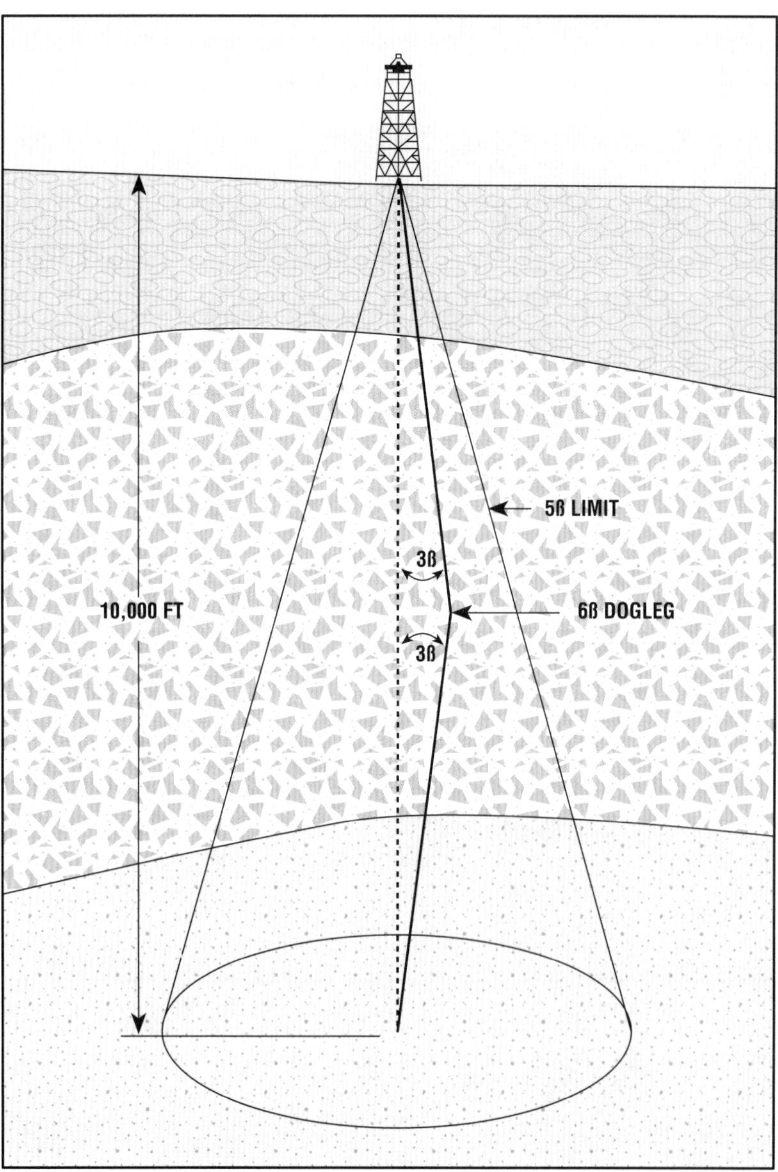

Figure 12. The typical 5-degree limit on total hole angle does not assure that a wellbore will be free of doglegs.

Restricted Rate of Hole-Angle Change

The most important consideration for drilling a usable hole is the rate of hole-angle change. By restricting the rate of hole-angle change, the operator prevents the occurrence of doglegs and keyseats.

Dogleg and Keyseats

As mentioned earlier, a dogleg is a sudden change in hole angle. Doglegs create problems for several reasons. One is that keyseats can form in doglegs. A keyseat forms because the drill string is in tension as it rotates through a dogleg. Being in tension, it tries to straighten itself as it rotates through the curve. If the lateral force is great enough, and the formation is soft enough, the drill pipe cuts into the formation and makes a groove the size of the tool joint. The groove is a keyseat. As the pipe rotates, it tends to cut a smaller groove which causes the next tool joint to ream through the narrower restriction. The tool joint's reaming the keyseat further slows drilling (fig. 13). What is more, when crew members pull the drill string out of the hole, the first drill collar, because it is usually larger in diameter than the keyseat, can get stuck in it. An expensive fishing job is often the result.

In general, shallow doglegs are more troublesome than those near the bottom of the hole because the drill string weight suspended below the dogleg increases as drilling progresses. The greater the suspended weight, the greater the tendency to cut a keyseat.

Operators determine the severity of a dogleg by the rate of hole-angle change within a certain length of hole drilled, usually 100 ft (30 m). If a 5-degree change in angle occurs in 100 ft (30 m) of drilled hole, the rate of dogleg is 5 degrees per 100 ft (30 m). If the 5-degree change in angle occurs in 50 ft (15 m) of drilled hole, the rate of dogleg is 10 degrees per 100 ft (30 m). A committee of the American Petroleum Institute (API) has developed a complete set of computerized tables for determining dogleg severity. These tables are available in API Bulletin D8, *A Tabular Method for Determining the Change of the Overall Angle and Dogleg Severity (for Hole Inclinations up to 70 Degrees).* This publication is available through API.

Problems severe doglegs and keyseats can cause include:

1. drill pipe fatigue;
2. stuck drill pipe, logging tools, and wirelines;
3. stuck, damaged, or worn production casing;
4. insufficient cementing circulation; and
5. production problems.

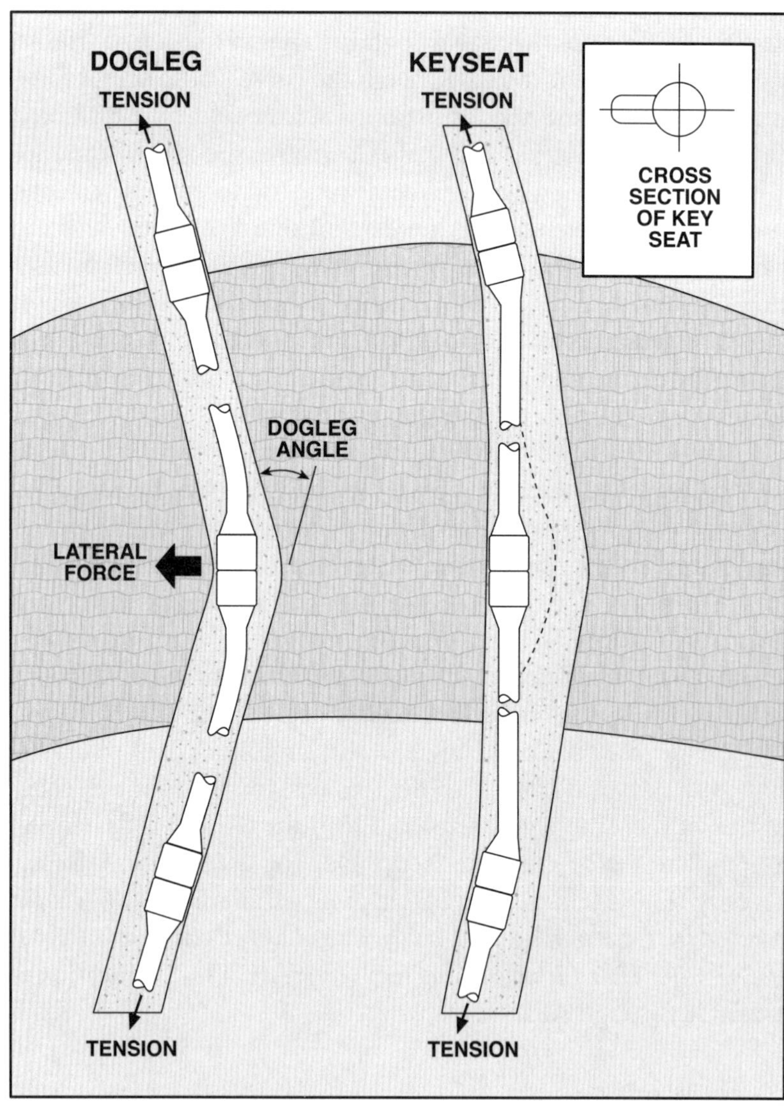

Figure 13. Formation of a keyseat. Continued drilling through a dogleg can result in the formation of a keyseat. When drill string is pulled out of the hole, large diameter drill collars can pull into the keyseat and become stuck.

Drill Pipe Fatigue

Drill pipe fatigue and damage result from rotating or pulling the drill string through a dogleg. When pipe is rotated through a dogleg, it undergoes cyclic variation in stress; each fiber is stressed alternately in tension and in compression. As the pipe rotates, the stress varies from maximum tension to minimum tension with each rotation (fig. 14). Adding tensile stress by adding more pipe to the string reduces the ability of the pipe to withstand cyclic, or bending, stress.

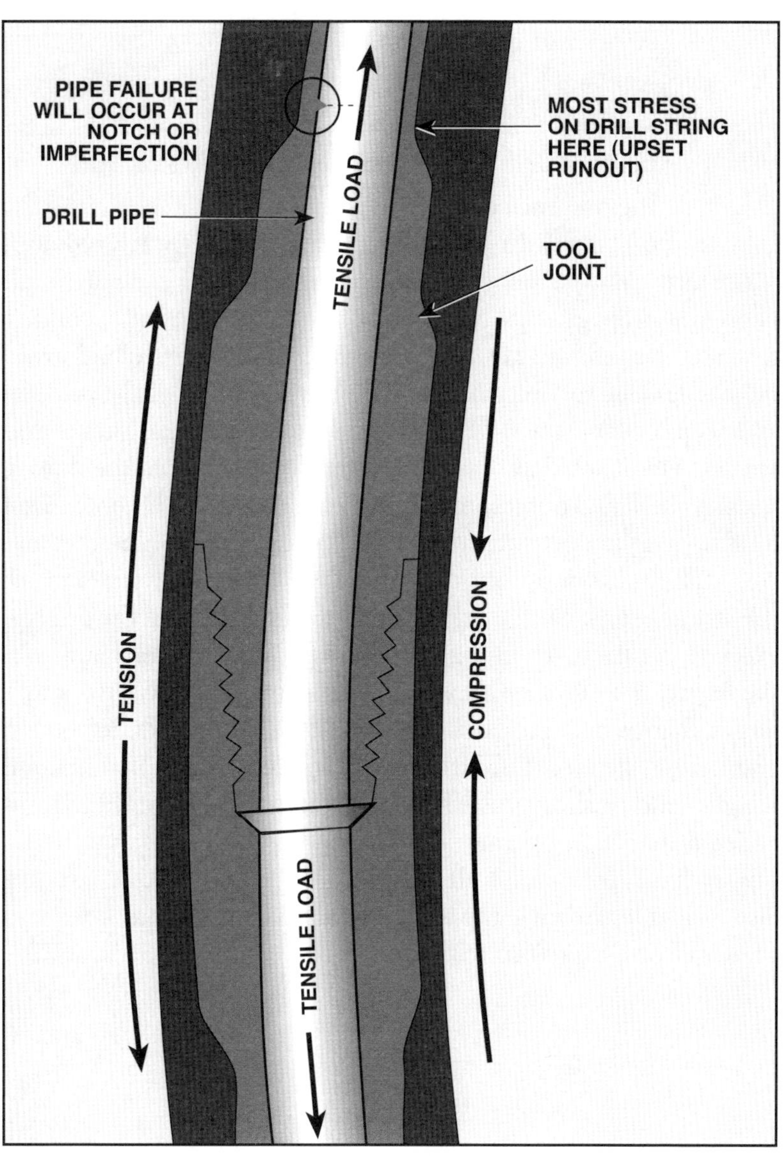

Figure 14. In a curved wellbore, alternating stresses are placed on the drill pipe as it rotates. The sharper the curve, the greater the tension and compression stress. High tensile load or pipe imperfections accelerate drill pipe failure.

If the bending stress exceeds the endurance limit of the metal, the pipe eventually fails because small irregularities grow into larger cracks and fractures. Most failures occur within the first 2 ft or 60 centimetres (cm) of the body adjacent to the tool joint; such failures can cause expensive fishing jobs or junked holes.

Factors affecting the degree of fatigue damage to the drill pipe include:

1. tensile load (weight) on the pipe at the dogleg;
2. severity of curvature of the dogleg;
3. mechanical dimension and physical properties of the pipe;
4. number of drill pipe rotations in the dogleg; and
5. the presence of corrosive drilling fluids.

In 1961, Arthur Lubinski, a research engineer, analyzed dogleg severity and published guidelines that showed the maximum degree of doglegs that drill pipe can tolerate without becoming damaged by fatigue. He believed that if contractors could avoid fatigue damage to pipe while drilling the hole, the hole would easily accept conventional designs of casing, tubing, and sucker rod strings. He concluded that the farther down the hole the dogleg occurs, the less troublesome. In the upper part of the hole, even mild doglegs are detrimental to drill collars, drill pipe, casing, and rods. The tension put on these tubulars by the heavy weight of the string below, combined with the bending stress in the dogleg, exceeds the tubular's endurance limits. On the other hand, if a dogleg is close to total depth, tension in the pipe is low, and a larger change in angle can be tolerated. Robert W. Nicholson, an engineer, expanded Lubinski's work to adjust for different mud weights and to show the effects of rotating off bottom. He published a chart that operators can use to determine maximum safe dogleg limits for grade E drill pipe (fig. 15). Simplified charts are also available showing the endurance limit for various sizes and grades of drill pipe (fig. 16).

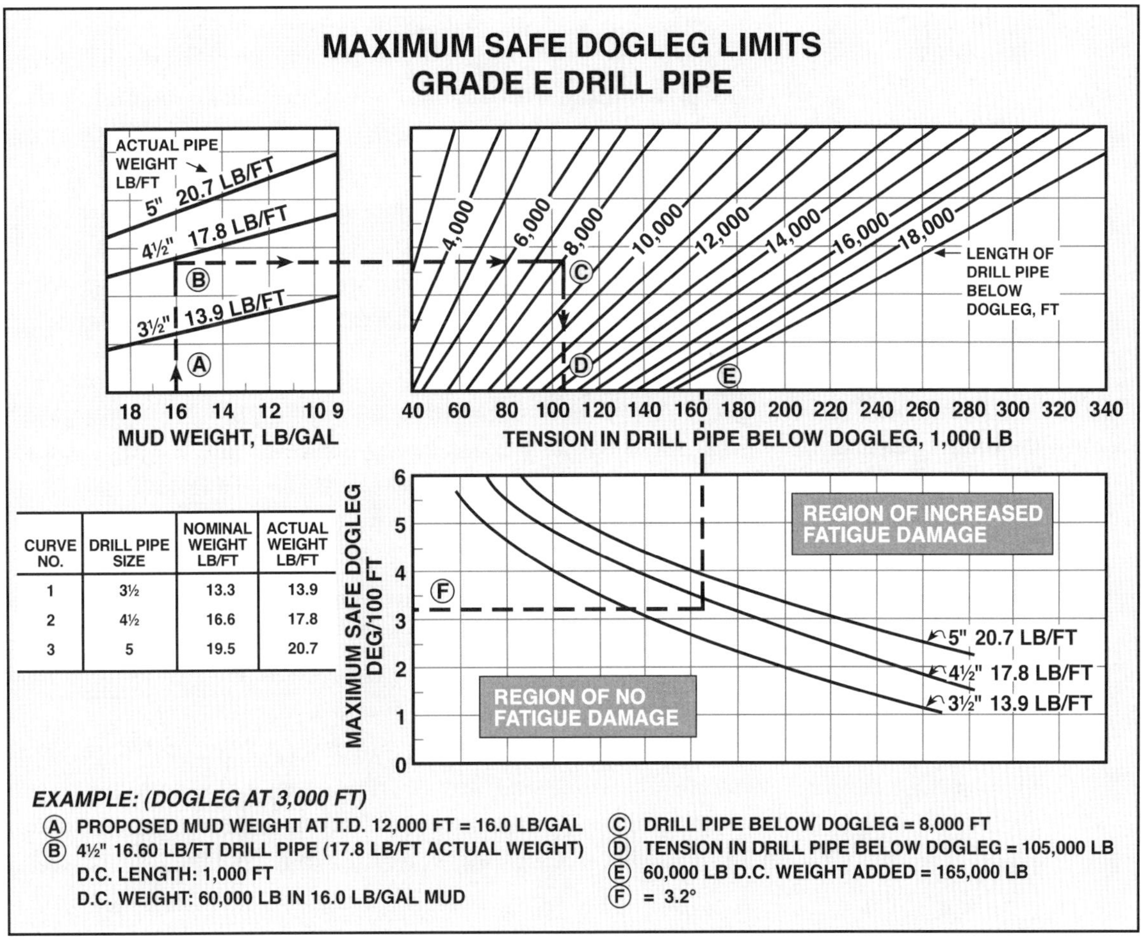

MAXIMUM SAFE DOGLEG LIMITS
GRADE E DRILL PIPE

ACTUAL PIPE WEIGHT LB/FT

5" 20.7 LB/FT

4½" 17.8 LB/FT

3½" 13.9 LB/FT

18 16 14 12 10 9

MUD WEIGHT, LB/GAL

4,000 6,000 8,000 10,000 12,000 14,000 16,000 18,000

LENGTH OF DRILL PIPE BELOW DOGLEG, FT

40 60 80 100 120 140 160 180 200 220 240 260 280 300 320 340

TENSION IN DRILL PIPE BELOW DOGLEG, 1,000 LB

CURVE NO.	DRILL PIPE SIZE	NOMINAL WEIGHT LB/FT	ACTUAL WEIGHT LB/FT
1	3½	13.3	13.9
2	4½	16.6	17.8
3	5	19.5	20.7

MAXIMUM SAFE DOGLEG DEG/100 FT

0 1 2 3 4 5 6

REGION OF INCREASED FATIGUE DAMAGE

REGION OF NO FATIGUE DAMAGE

5" 20.7 LB/FT

4½" 17.8 LB/FT

3½" 13.9 LB/FT

EXAMPLE: (DOGLEG AT 3,000 FT)

- Ⓐ PROPOSED MUD WEIGHT AT T.D. 12,000 FT = 16.0 LB/GAL
- Ⓑ 4½" 16.60 LB/FT DRILL PIPE (17.8 LB/FT ACTUAL WEIGHT)
  D.C. LENGTH: 1,000 FT
  D.C. WEIGHT: 60,000 LB IN 16.0 LB/GAL MUD
- Ⓒ DRILL PIPE BELOW DOGLEG = 8,000 FT
- Ⓓ TENSION IN DRILL PIPE BELOW DOGLEG = 105,000 LB
- Ⓔ 60,000 LB D.C. WEIGHT ADDED = 165,000 LB
- Ⓕ = 3.2°

Figure 15. Maximum dogleg limits for grade E drill pipe (Courtesy of Oil and Gas Journal and Drilco)

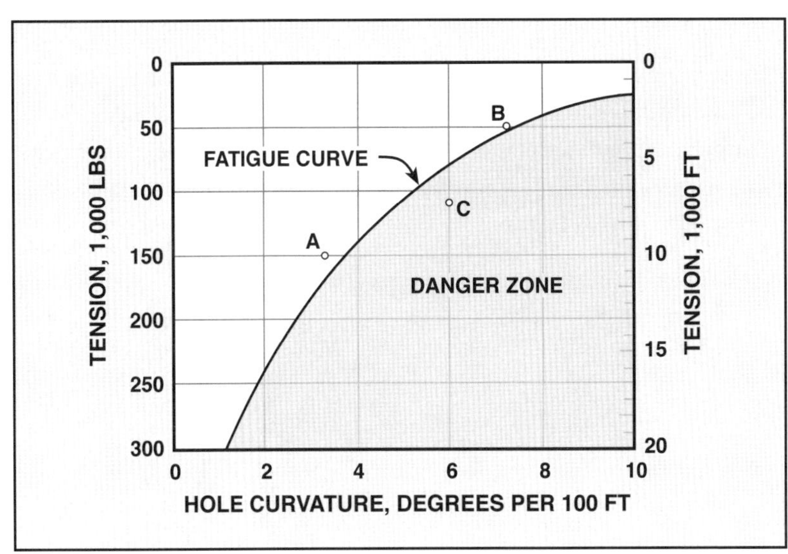

Figure 16. Chart showing the endurance limit for 4½-inch (114.3-millimetres), 16.6 ppf, grade E drill pipe in a gradual dogleg. Damage to drill pipe is avoided when conditions fall to the left of the fatigue curve (points A, B) but drill pipe failure is likely if conditions fall to the right of the fatigue curve (C). (Courtesy of Smith International)

Stuck Drill Pipe, Logging Tools, and Wireline

In addition to causing drill pipe fatigue, doglegs increase the possibility that the drill string will become stuck in the wellbore. When pipe is rotated through a dogleg, the hole may slough, which may cause the pipe to stick. Pipe may also become stuck when the driller pulls large-diameter drill collars into a keyseat when tripping out of the hole. In addition, logging tools and wireline may become stuck or lost in keyseats and must be fished or milled out. Various actions the crew takes to correct doglegs can also damage the wall of the hole.

Stuck, Damaged, or Worn Production Casing

Operators drill wellbores with the intention of setting casing and completing the hole as a producing well. Running casing through a dogleg can cause serious problems. If the casing becomes stuck in the dogleg, it may not extend through the productive zone. The crew will then have to drill out the shoe and set a smaller-size casing set through the productive zone. Even if the crew ran the casing through the dogleg to the bottom, the dogleg could damage the casing so badly that the operator could not run production equipment in the hole.

During drilling operations, the lateral force of drill pipe rotating inside casing set through the dogleg, or dragging through it while tripping, can wear a hole through the casing and cause drilling problems, possibly even a blowout.

Insufficient Cement Circulation

Casing run through a dogleg is forced tightly against the wall of the hole. Because no cement can circulate between the wall of the hole and the casing at this point, a good cement bond is not possible.

Production Problems

Efficient production requires a smooth string of casing. Doglegs can cause sucker rod wear and tubing leaks that lead to expensive repair jobs. Distorted or collapsed casing can also cause packers and tools to become stuck when they are run in and out of the well.

To summarize—

Definition of a straight hole

- a hole that stays within the boundary of a cone, as designated by the operator in the deviation clause of the contract
- a hole that does not change direction rapidly; usually, no more than 3 degrees per 100 ft (30 m) of hole

Reasons to restrict total hole angle

- to ensure hole bottoms out within a reasonable distance from surface
- to reduce costs by not drilling extra footage incurred by a crooked hole
- to prevent legal disputes by keeping wellbore within lease boundaries
- to keep wellbores properly spaced for a particular lease
- to assure that the wellbore penetrates a specific pay zone
- to ensure that the wellbore does not drift to a point below an oil-gas-water contact

Doglegs and keyseats

- a dogleg is a sudden change in hole direction
- a keyseat is a groove worn into the side of the hole; it is caused by tool joints and drill pipe coming in contact with the side of the hole in a keyseat

Dogleg and keyseat problems

- drill pipe fatigue—results from rotating or pulling string through a dogleg
- stuck pipe—when drill string is pulled, drill collar string can stick in keyseat; or formation may slough during drilling, which may cause pipe to stick
- stuck, damaged, or worn casing—casing can become stuck, damaged, or worn when run through a dogleg

- insufficient circulation of cement—casing is forced against side of dogleg so no cement can circulate between wall of hole and casing; a good cement bond fails to occur
- production problems—doglegs can wear sucker rods and cause tubing leaks

---

Factors Affecting Hole Deviation

A hole may deviate from the vertical for a number of reasons. Both the formations being drilled and the equipment used to drill them have decided effects on hole deviation.

Formation Effects

The type of formation being drilled affects hole deviation. Two formation characteristics that affect the bit's ability to drill a straight hole are formation dip and the variation in the drillability of formations.

Formation Dip

Formation dip refers to the degree of inclination of the bedding planes of the formation. In formations with dips of 40 degrees or less, the bit tends to drill (walk) updip, or into the hill (fig. 17). Updip drilling tends to be more prevalent in laminar formations, such as interbedded shale and sandstone, than in thick, uniform formations, such as limestone.

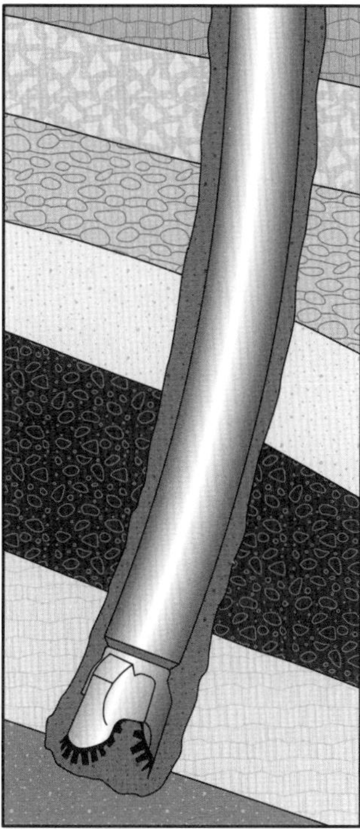

Figure 17. In laminar formations with dips of 40 degrees or less, the bit tends to drill (walk) updip.

In formations with dips of 40 degrees or more, or severely inclined formations, the bit tends to drill parallel to the bedding planes because the formations fracture easily along bedding planes (fig. 18). The bit thus tends to follow the line of least resistance and slide downdip.

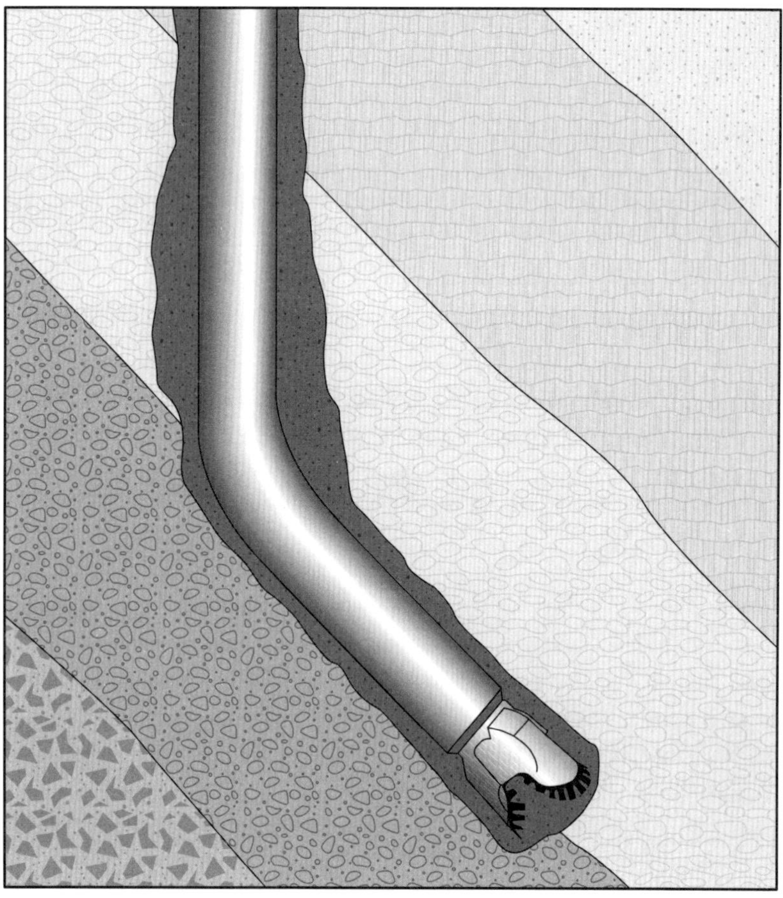

Figure 18. In laminar formations with dips of 40 degrees or more, the bit follows the line of least resistance in the soft layers and tends to drill (slide) downdip parallel to the bedding planes.

Variation in Drillability of Formations

The drillability, or drilling rate, of rocks depends on the compressive strength of the rock, a factor that varies over extremely wide ranges (fig. 19). Generally, a bit drills slower in rocks with high compressive strength (hard rocks) than in rocks with low compressive strength (soft rocks). When a bit drills in formations that alternate from hard to soft, this hard-to-soft alternation deflects the bit from a normal course. The soft sections wash out, and since the drill collars are smaller in diameter than the bit, the bit moves laterally within the soft formations. Then, as the bit drills through the next hard formation, the soft formation above has an offset ledge. After the bit has drilled through several alternating hard and soft sections, the hole through the hard sections may not be in line (fig. 20). This ledge effect is more pronounced in dipping formations.

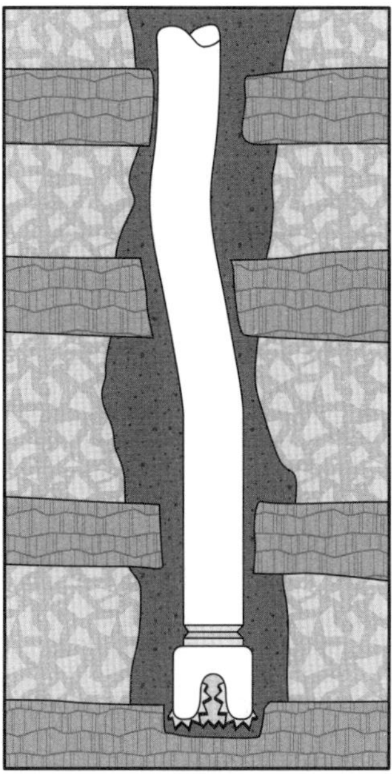

Figure 20. Drilling through alternating hard and soft ledges can produce offset ledges. (Courtesy of Smith International, Inc.)

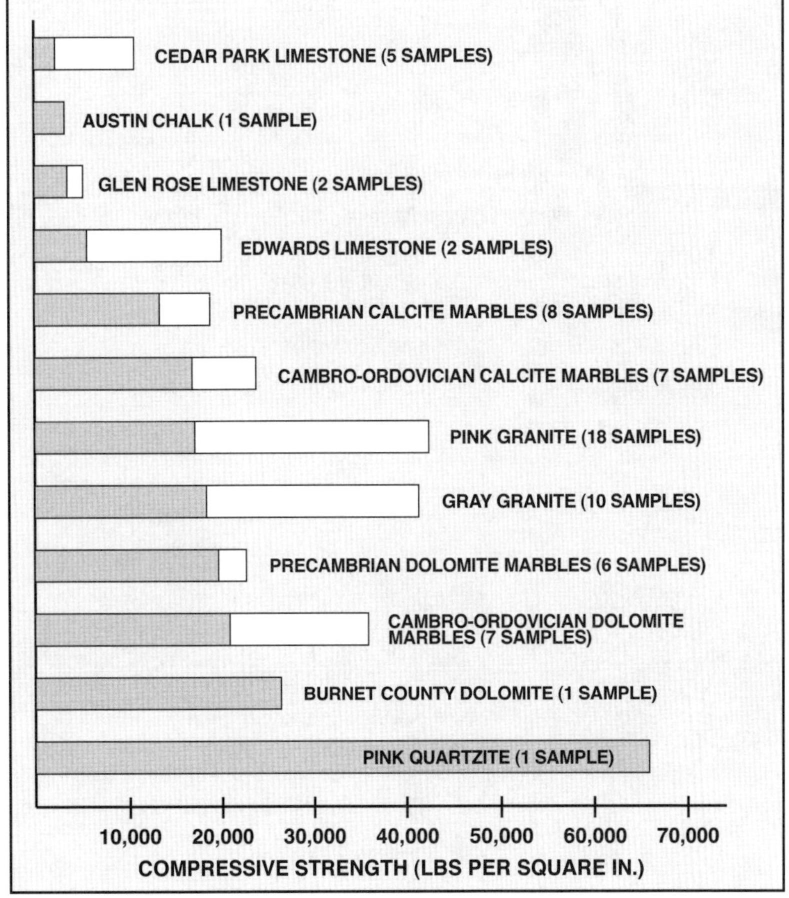

Figure 19. Compressive strength of rocks. Compressive strength denotes weight in pounds-per-square-inch required to drill rock with a three-cone bit. The unshaded area indicates a range of compressive strength from multiple samples.

Mechanical Effects

In addition to the formation effects that cause holes to deviate, the equipment used when drilling those formations can also cause crooked holes.

Drill strings bend. They do not remain perfectly straight. With no weight on the bit, gravity is the only force acting on the drill string. Gravity tends to bring the hole toward vertical (fig. 21).

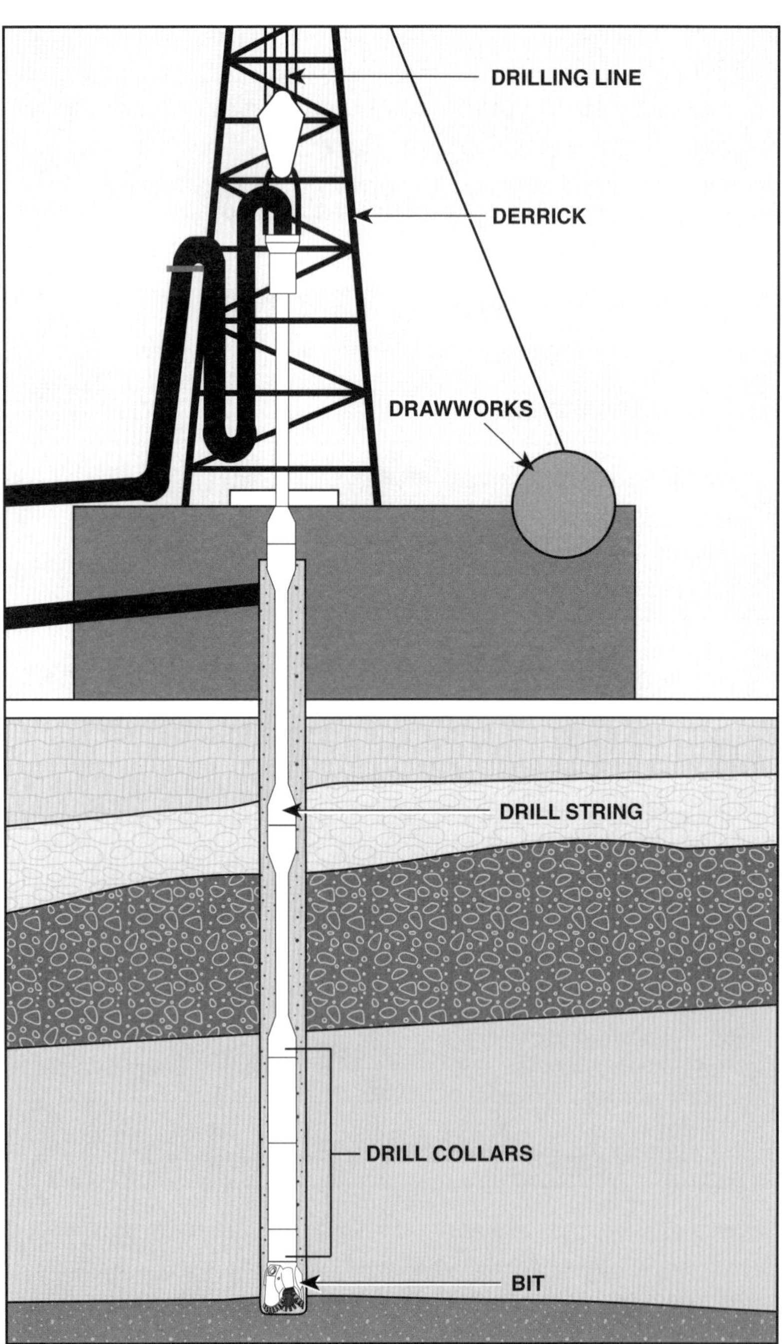

Figure 21. The drill string is kept in tension by two opposing forces—the weight of the drill collars and the pull of the drawworks and drilling line.

To make hole, however, the driller must apply the weight of the drill collars to the drill bit. However, the more weight the driller adds, the greater is the tendency of the drill string to bend and buckle. The bending can cause crooked hole (fig. 22). In nondipping formations, a limber, unstable bottomhole assembly (BHA) can also cause a crooked hole. A limber BHA allows the bit to be easily deflected and to wobble off-center, which results in a spiral borehole (fig. 23).

If a rig uses undersized drill collars, the excessive clearance between the collars and the wall of the hole makes it possible for the bit to move laterally. This lateral movement of the bit can cause a crooked hole. A stabilized bit drills longer, faster, and straighter than a bit that is allowed to wobble and rotate off-center.

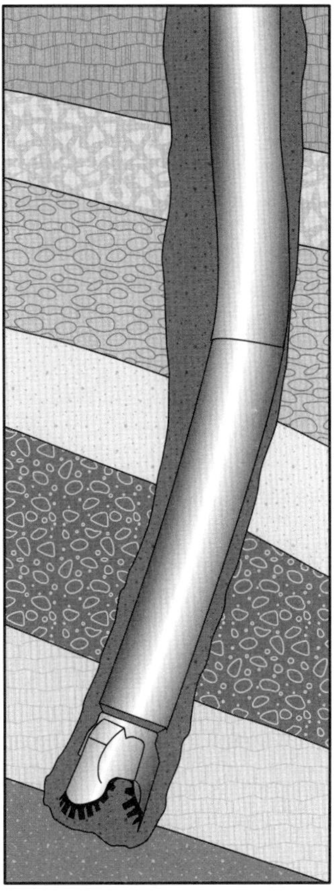

Figure 22. The bending of a drill string will cause the bit to deviate and drill a crooked hole.

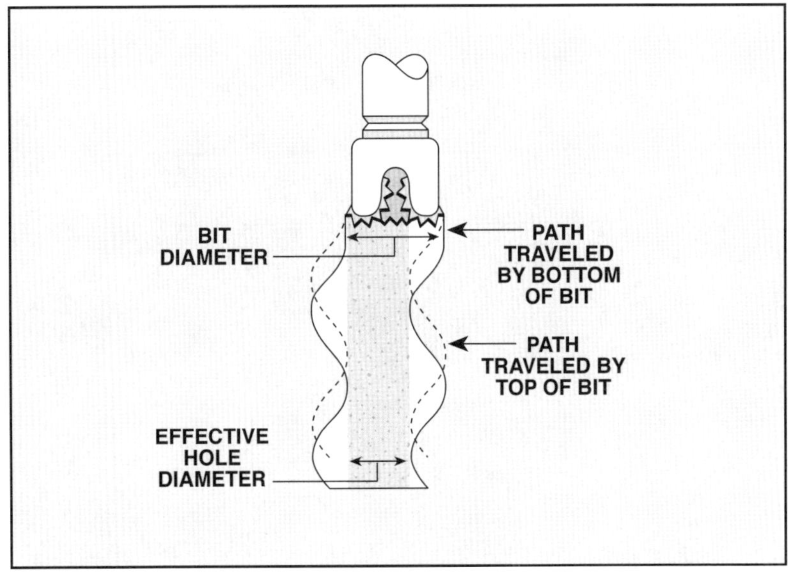

Figure 23. A spiral hole caused by drilling with an unstabilized bit in nondipping formations. Note that the effective hole diameter is less than actual bit size. (Courtesy of Smith International, Inc.)

Dull or balled-up bits can also cause crooked holes; both conditions require more weight on bit (WOB) to maintain a satisfactory rate of penetration or ROP (fig. 24). More weight on bit tends to bend the drill collar string, which forces the bit away from vertical. More weight on bit may also be needed if the bit balls up. Drilling fluid cannot readily exit a balled-up bit, resulting in inadequate fluid volume. Inadequate fluid volume leads to reduced rate of penetration because reduced jet velocity does not remove cuttings efficiently. Thus, the bit redrills cuttings and not uncut formation.

Figure 24. Bits affect straight hole drilling. Dull bits require more weight and increase the tendency to deviate from vertical. (Courtesy of Hughes Christensen)

Bit records from previous wells should be evaluated during well planning because bit selection can influence hole deviation. For example, a bit with no cone offset tends to drill straighter than one with cone offset. Further, fixed-cutter such as natural diamond bits and polycrystalline diamond compact (PDC) bits drill straighter than roller cone bits (fig. 25A, B). Hammer bits (fig. 25C) are used in air

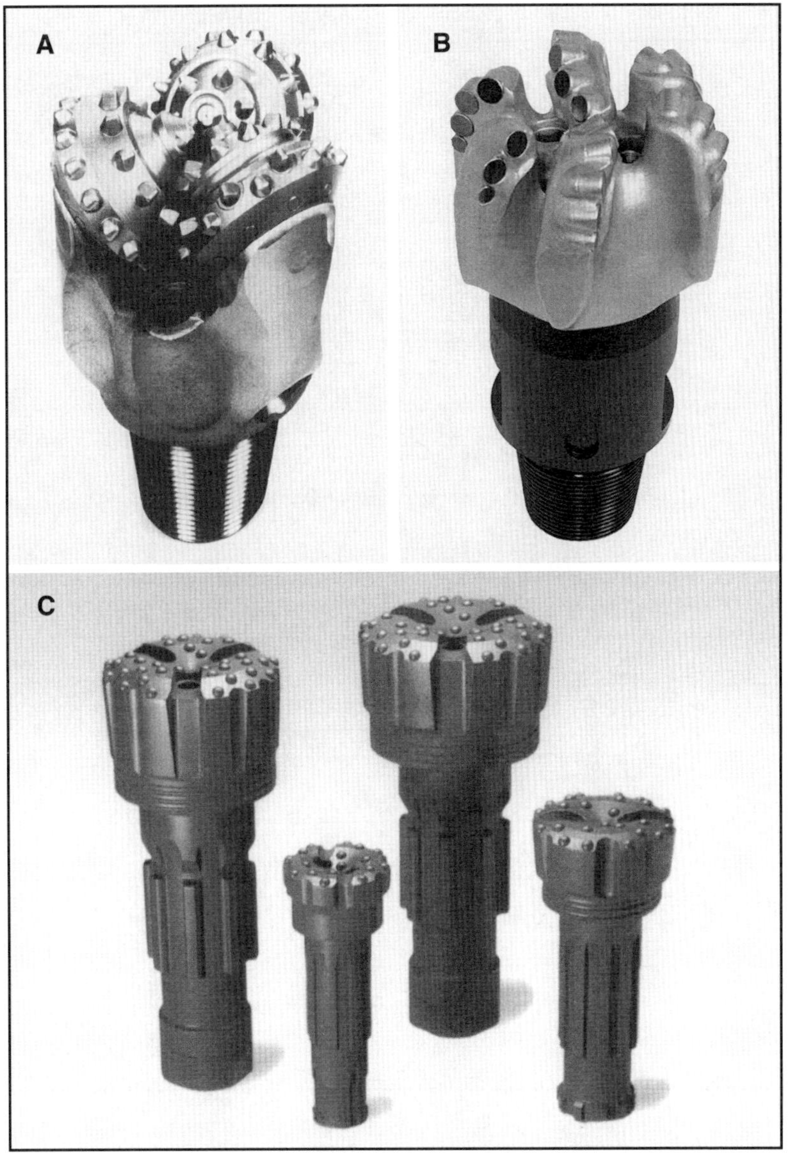

Figure 25. Bit selection affects straight hole drilling. A. Offset cones increase the tendency to wobble (Courtesy of Reed Hycalog); B. Fixed cutter bits (PDC) drill straighter holes; C. Hammer bits can drill straighter holes than conventional bits.(Courtesy of Hughes Christensen)

percussion tools (fig. 26) in some hardrock basins. The tool-and-bit combination operates like a jackhammer and has demonstrated the ability to maintain a straighter hole than conventional rotary bits in difficult, hard rock, crooked-hole country (fig. 27).

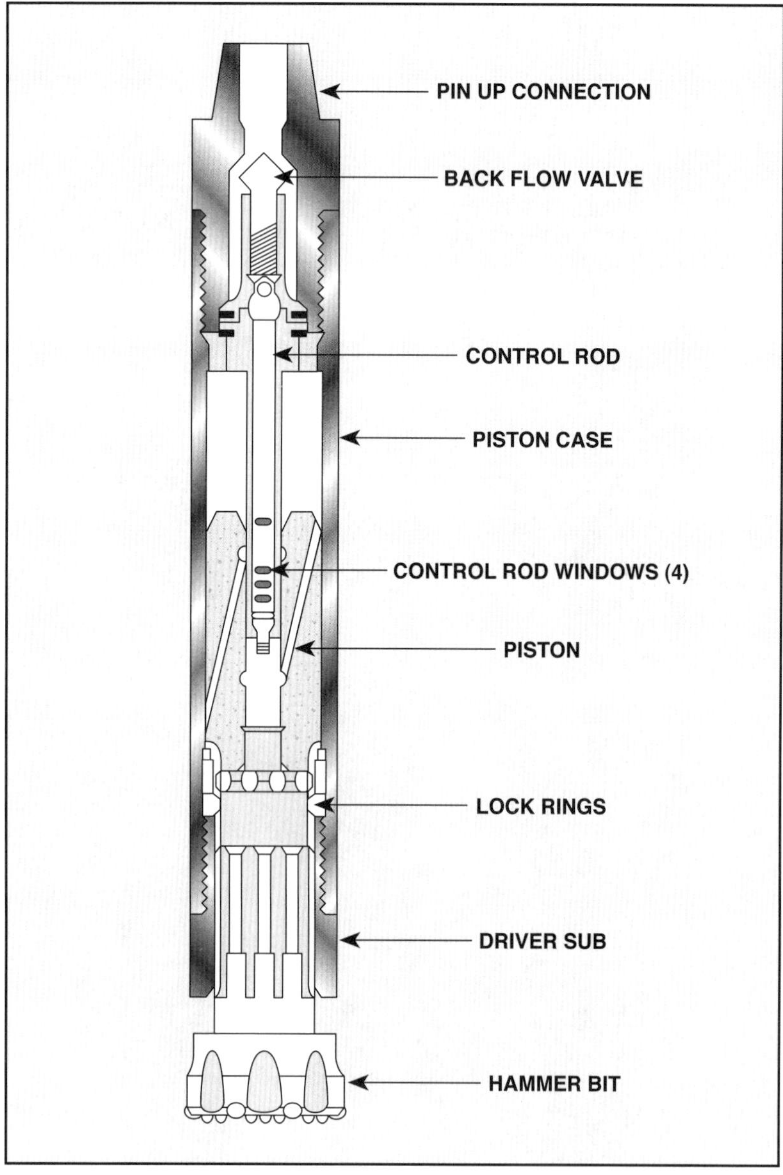

Figure 26. Air-hammer (percussion) drilling tools (Courtesy of Smith Bits Division of Smith International, Inc.)

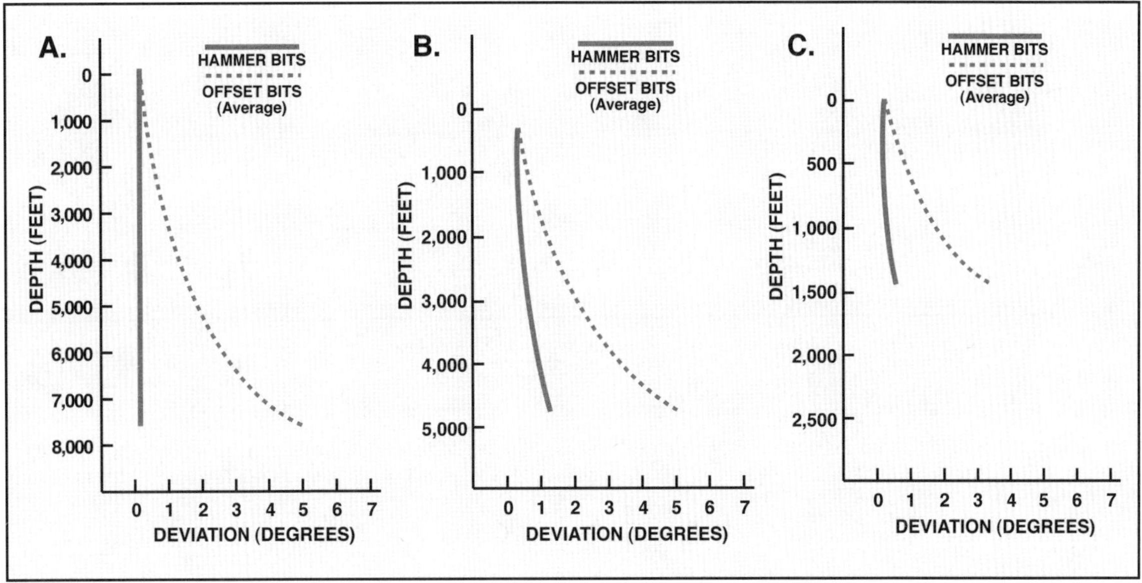

Figure 27. Comparison of hole deviation in control wells using hammer bits versus offset wells using rotary bits in hard rock country; A. Appalachian Basin; B. Arkoma Basin; C. Permian Basin. (Courtesy of Society of Petroleum Engineers)

To summarize—

Factors affecting hole deviation

- formation dip—in dips of 40 degrees or less, hole goes updip; in dips of 40 degrees or more, hole goes down dip (follows the bedding planes)
- variation in formation's drillability—in beds that alternate soft to hard or hard to soft, bit is deflected from normal course
- mechanical effects—drill strings bend, which affects hole direction; and dull or balled-up bits can drill crooked hole if WOB is increased

---

Methods of Controlling Hole Deviation

Although crew members can sometimes correct a crooked hole, the operation is often expensive. They may have to plug back and redrill a badly crooked portion of the hole to straighten it, or, in extreme cases, they may have to skid the rig and start a new hole. The best approach to use when drilling in crooked-hole country is to employ preventive measures. These measures include drilling a shallow and vertical surface hole, selecting the best drilling method for the area—for example, standard rotary, air rotary, air percussion, or downhole mud motor—an appropriate BHA, and reducing weight on bit as much as possible. Bit records, deviation surveys, drilling time charts, daily drilling reports (fig. 28), a geological prognosis, and information from nearby wells can provide valuable well-planning assistance.

CODE NO. - OPERATION	MORN.	DAY	EVE.
TIME DISTRIBUTION – HOURS			
1. RIG UP AND TEAR DOWN			
2. DRILL ACTUAL			
3. REAMING			
4. CORING			
5. CONDITION MUD & CIRCULATE			
6. TRIPS			
7. LUBRICATE RIG			
8. REPAIR RIG			
9. CUT OFF DRILLING LINE			
10. DEVIATION SURVEY			
11. WIRE LINE LOGS			
12. RUN CASING & CEMENT			
13. WAIT ON CEMENT			
14. NIPPLE UP B.O.P.			
15. TEST B.O.P.			
16. DRILL STEM TEST			
17. PLUG BACK			
18. SQUEEZE CEMENT			
19. FISHING			
20. DIR. WORK			
21.			
22.			
23.			
COMPLETION: A. PERFORATING			
COMPLETION: B. TUBING TRIPS			
COMPLETION: C. TREATING			
COMPLETION: D. SWABBING			
COMPLETION: E. TESTING			
COMPLETION: F.			
COMPLETION: G.			
COMPLETION: H.			
TOTALS			
DAYWORK TIME SUMMARY (OFFICE USE ONLY)			
HOURS W/CONTR. D.P.			
HOURS W/OPR. D.P.			
HOURS WITHOUT D.P.			
HOURS STANDBY			
TOTAL DAYWORK			
NO. OF DAYS FROM SPUD			
CUMULATIVE ROTATING HOURS			
DAILY MUD COST			
TOTAL MUD COST			

MORNING TOUR

DRILLING ASSEMBLY (At end of tour)

NO.	ITEM	LENGTH
	BIT	
	OD	
	OD	
	OD	
	OD	
	OD	
	STANDS ___ D.P.	
	SINGLES ___ D.P.	
	KELLY DOWN	
	TOTAL	
WT. OF STRING		

BIT RECORD

BIT NO.		
SIZE		
IADC CODE		
MANUFACTURER		
TYPE		
SERIAL NO.		
JETS		
TFA		
DEPTH OUT		
DEPTH IN		
TOTAL DRILLED		
TOTAL HOURS		

CUTTING STRUCTURE: INNER	OUTER	DULL CHAR.	LOCATION
BEARINGS/SEALS	GAGE	OTHER DULL CHAR.	REASON PULLED

MUD RECORD

TIME			
WEIGHT			
PRESSURE GRADIENT			
FUNNEL VISCOSITY			
PV/YP	/	/	/
GEL STRENGTH	/	/	/
FLUID LOSS			
pH			
SOLIDS			

MUD & CHEMICALS ADDED: TYPE	AMOUNT	TYPE	AMOUNT

DAY TOUR

DRILLING ASSEMBLY (At end of tour)

NO.	ITEM	LENGTH
	BIT	
	OD	
	OD	
	OD	
	OD	
	OD	
	STANDS ___ D.P.	
	SINGLES ___ D.P.	
	KELLY DOWN	
	TOTAL	
WT. OF STRING		

BIT RECORD

BIT NO.		
SIZE		
IADC CODE		
MANUFACTURER		
TYPE		
SERIAL NO.		
JETS		
TFA		
DEPTH OUT		
DEPTH IN		
TOTAL DRILLED		
TOTAL HOURS		

CUTTING STRUCTURE: INNER	OUTER	DULL CHAR.	LOCATION
BEARINGS/SEALS	GAGE	OTHER DULL CHAR.	REASON PULLED

MUD RECORD

TIME			
WEIGHT			
PRESSURE GRADIENT			
FUNNEL VISCOSITY			
PV/YP	/	/	/
GEL STRENGTH	/	/	/
FLUID LOSS			
pH			
SOLIDS			

MUD & CHEMICALS ADDED: TYPE	AMOUNT	TYPE	AMOUNT

EVENING TOUR

DRILLING ASSEMBLY (At end of tour)

NO.	ITEM	LENGTH
	BIT	
	OD	
	OD	
	OD	
	OD	
	OD	
	STANDS ___ D.P.	
	SINGLES ___ D.P.	
	KELLY DOWN	
	TOTAL	
WT. OF STRING		

BIT RECORD

BIT NO.		
SIZE		
IADC CODE		
MANUFACTURER		
TYPE		
SERIAL NO.		
JETS		
TFA		
DEPTH OUT		
DEPTH IN		
TOTAL DRILLED		
TOTAL HOURS		

CUTTING STRUCTURE: INNER	OUTER	DULL CHAR.	LOCATION
BEARINGS/SEALS	GAGE	OTHER DULL CHAR.	REASON PULLED

MUD RECORD

TIME			
WEIGHT			
PRESSURE GRADIENT			
FUNNEL VISCOSITY			
PV/YP	/	/	/
GEL STRENGTH	/	/	/
FLUID LOSS			
pH			
SOLIDS			

MUD & CHEMICALS ADDED: TYPE	AMOUNT	TYPE	AMOUNT

IADC - API OFFICIAL DAILY DRILLING REPORT FORM

APPROVED — PRINTED IN U.S.A. — APPROVED

Figure 28. Rig site daily drilling report form. Note that column 2 is a description of the drilling assembly to be completed by each tour. (Courtesy of IADC)

Start Straight

A vertical start in the upper portion of the hole averts many problems in maintaining a straight hole later. When hard rocks lie at or near the surface, the driller may have difficulty starting and maintaining a vertical hole. Keeping a surface hole straight is even more difficult if the hard rocks are inclined, because, given that the driller can only apply light weight on bit near the surface, the bit has a great tendency to walk updip.

Modern drilling procedures using downhole motors, PDC bits, and MWD, allow much tighter control of hole deviation. In crooked-hole country, operators can utilize the capabilities provided by these drilling technologies to assure a straight hole or to correct excess deviation should it occur. With downhole motors, the drill string does not rotate; thus, vibration is reduced, bit stability is greatly enhanced, and the tendency to deviate from the vertical is reduced (figs. 29, 30, 31).

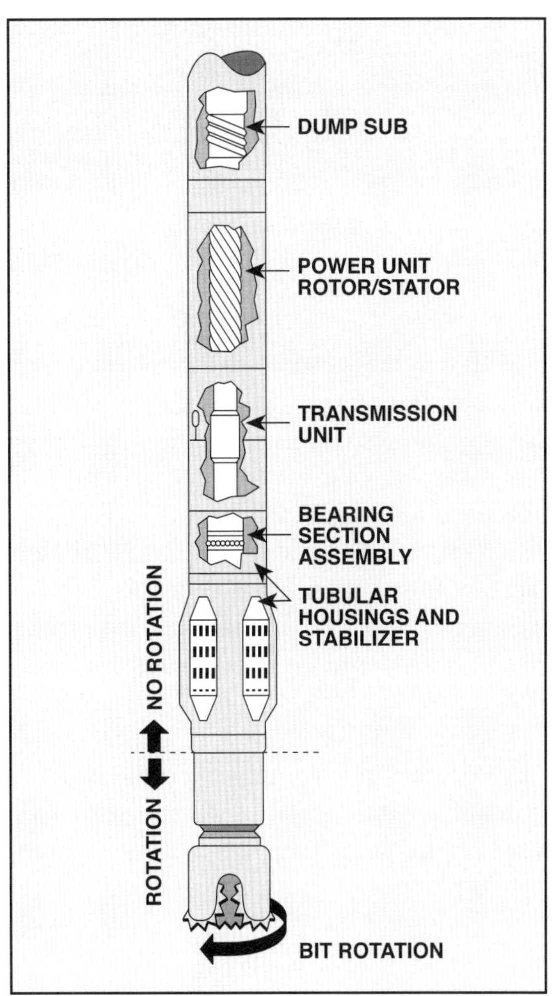

Figure 29. A downhole mud motor reduces vibration and bit movement because the drill string above the bit is not rotating. (Courtesy of Sperry-Sun Drilling Services, a Halliburton Company)

Figure 30. A downhole mud motor rotor, or stator (at left) and power unit housing

Figure 31. Downhole mud motors in the shop ready for moving to the field

Air drilling with standard rotary methods has proven effective for fast straight-hole drilling in hard rock areas like the Arkoma, Val Verde, and Appalachian Basins (fig. 32). With air drilling, the air or gas cools the bit so effectively that operators can apply higher-than-normal weight on bit, which makes for high ROP and less tendency for the bit to drill off-vertical.

Air percussion tools (see fig. 26) can also solve hole deviation problems in shallow surface formations. Crew members install a percussion tool (an air hammer) above the bit and run it to bottom. Circulating air or gas creates a hammering action (causes the tool to rapidly move up and down) on the bit and, when this hammering effect is combined with rotation, ROP is rapid. Also, percussion tools do not require a great deal of weight on bit to operate effectively. Thus, an air hammer tool not only drills straighter (see fig. 26), but also faster because its drilling action is not wholly dependent on weight on bit.

Figure 32. Skid-mounted air compressors for an air-rotary drilling rig

Two other methods are available to drill a straight and vertical surface hole where the surface formations are hard. One method involves using a BHA that creates a pendulum effect (the pendulum effect is discussed in the following section). The other method involves first drilling a pilot hole, which is a hole that is smaller in diameter than the diameter of the final surface hole. Then, the rig uses a hole opener to ream the pilot hole to the required diameter for surface casing. The operator can drill a vertical small-diameter pilot hole much easier than a full-gauge hole because a small-diameter bit does not require as much weight to drill straight as a large-diameter bit.

Use the Right Bottomhole Assembly

The bottomhole assembly, or BHA, is the portion of the drill string below the drill pipe. The primary consideration affecting selection of a drilling assembly is the nature of the formations to be penetrated. Operators must determine the crooked-hole tendencies of the formations by evaluating data on formation thickness, dip, faulting, fracturing, firmness, and drillability. Then, they can classify each formation as hard, medium-hard, medium soft, or soft, and as either abrasive or nonabrasive.

The hole encounters different formations at various depths. In most cases, operators can predict the characteristics of these formations before they drill them. For example, they usually have a geological prognosis that predicts the depth for each formation to be penetrated. They also obtain data for each hole size to be drilled from records of wells previously drilled in the area, geological maps, and seismic surveys. Finally, they can classify each interval of hole as having (1) mild, (2) medium, or (3) severe crooked-hole tendencies.

Once operators have classified the hole intervals, they can select the size of hole they wish to drill in each interval and the type of BHA to use in each interval. BHAs vary from simple to complex. A simple BHA may be slick—that is, only a bit and regular drill collars compose the BHA. Complex BHAs, on the other hand, may consist of several special drill collars, various types of stabilizers and reamers, and a special bit.

Four types of BHAs to control deviation include:

1. the pendulum assembly;
2. the packed-hole assembly ;
3. the packed pendulum assembly; and
4. various combinations of downhole motors, MWD, and steerable assemblies.

Pendulum Assembly

When a driller is making hole with a slick string of drill pipe and collars, he has only one thing going for him to keep the hole on a vertical course: the pendulum effect. The pendulum effect is the tendency of the drill string to hang in a vertical position because of the force of gravity (see fig. 21). If a hole deviates from vertical, the bit and the drill collars lie on the low side of the hole and seek to return to vertical unless an opposing force prevents them from doing so (fig. 33).

On the bottom portion of the drill string, three forces seek to restore the pendulum to a vertical position in an inclined hole: (1) gravity; (2) the axial, or drill collar, load; and (3) the reaction of the formation to these loads (see fig. 33).

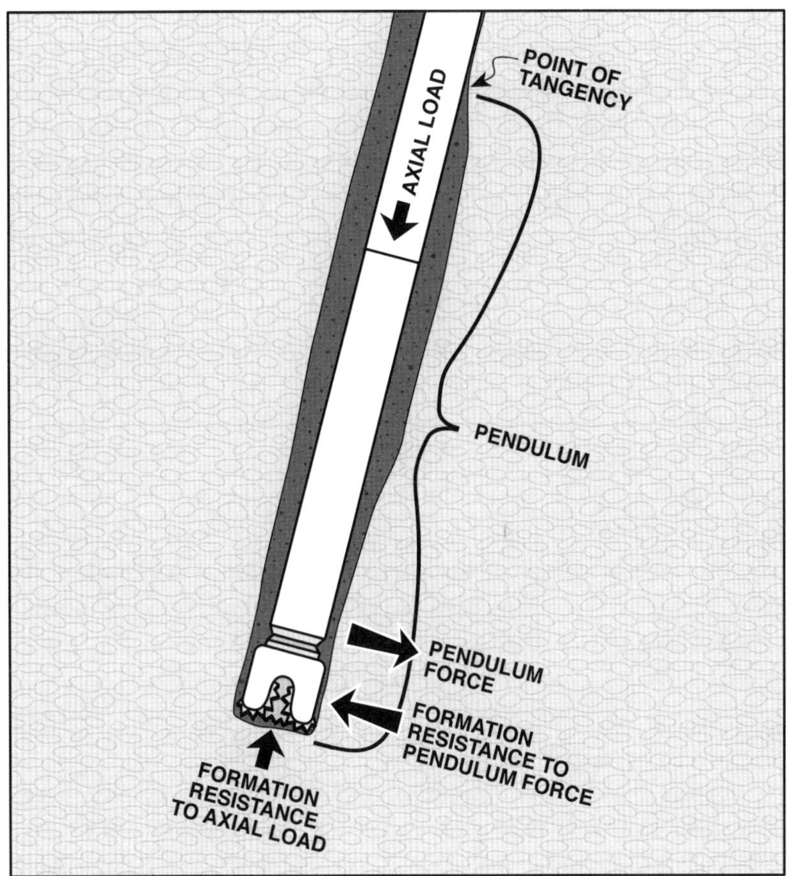

Figure 33. Pendulum forces at work. In an inclined hole the bit and drill collars lie on the low side of the hole. Below the point of tangency, gravity exerts a pendulum force that seeks to restore the drill collars and bit to vertical.

The weight of the drill collar string between the bit and the first point of contact the string makes with the wall of the hole—the point of tangency—supplies the pendulum force. The higher the point of tangency, the longer the pendulum, and the stronger is its tendency to return to vertical. The weight of the drill collar string supplies the axial load, and axial load influences the pendulum force. A greater load, or lighter drill collars, causes the bottom of the string to buckle closer to the bit. Buckling closer to the bit lowers the point of tangency and reduces pendulum force.

The formation opposes axial load and pendulum force. The formation reaction is a combination of two forces, one parallel to the axis of the hole, and one perpendicular to the axis of the hole.

When pendulum force equals formation reaction in an inclined hole, a state of equilibrium exists, and the hole continues straight, but inclined. If pendulum force is greater, the hole angle decreases. If the formation reaction is greater, hole angle increases.

A pendulum BHA works on the principle of the pendulum effect. One pendulum BHA is slick—that is, it is composed of a bit and several large-diameter drill collars. Another pendulum assembly has one or more stabilizers installed in the drill collar string in a predetermined position above the bit.

Arthur Lubinski and Henry Woods designed the pendulum assembly and made mathematical calculations to show proper placement of the first stabilizer for maximum pendulum force. Operators install the stabilizer as high as possible above the bit without allowing the drill collars between it and the bit to touch the wall of the hole. The first stabilizer becomes the point of tangency and acts as a fulcrum point (fig. 34). Its placement is related to weight on bit, hole size, drill collar size, formation drillability, and formation dip. Operators can obtain tables, graphs, and computer software programs showing where to place the fulcrum stabilizer that controls deviation under many conditions. Tables, such as Table 1, are available for most common hole sizes and formation dips, up to 45 degrees.

Assume you are drilling an 8¾-inch (in.) or a 222.25-millimetre (mm) hole with 6-in. (152.4-mm) drill collars in a uniform formation with a known dip of 15 degrees. You have maintained a hole angle of 4 degrees for about 100 ft (30.5 m) with 13,237 pounds (lb) or 5,850.75 decanewtons (dN) weight on bit. You have established an equilibrium condition. However, penetration rates are slow. What changes could you make in the BHA that would allow greater weight on bit and an increased penetration rate but not exceed the 4-degree hole angle?

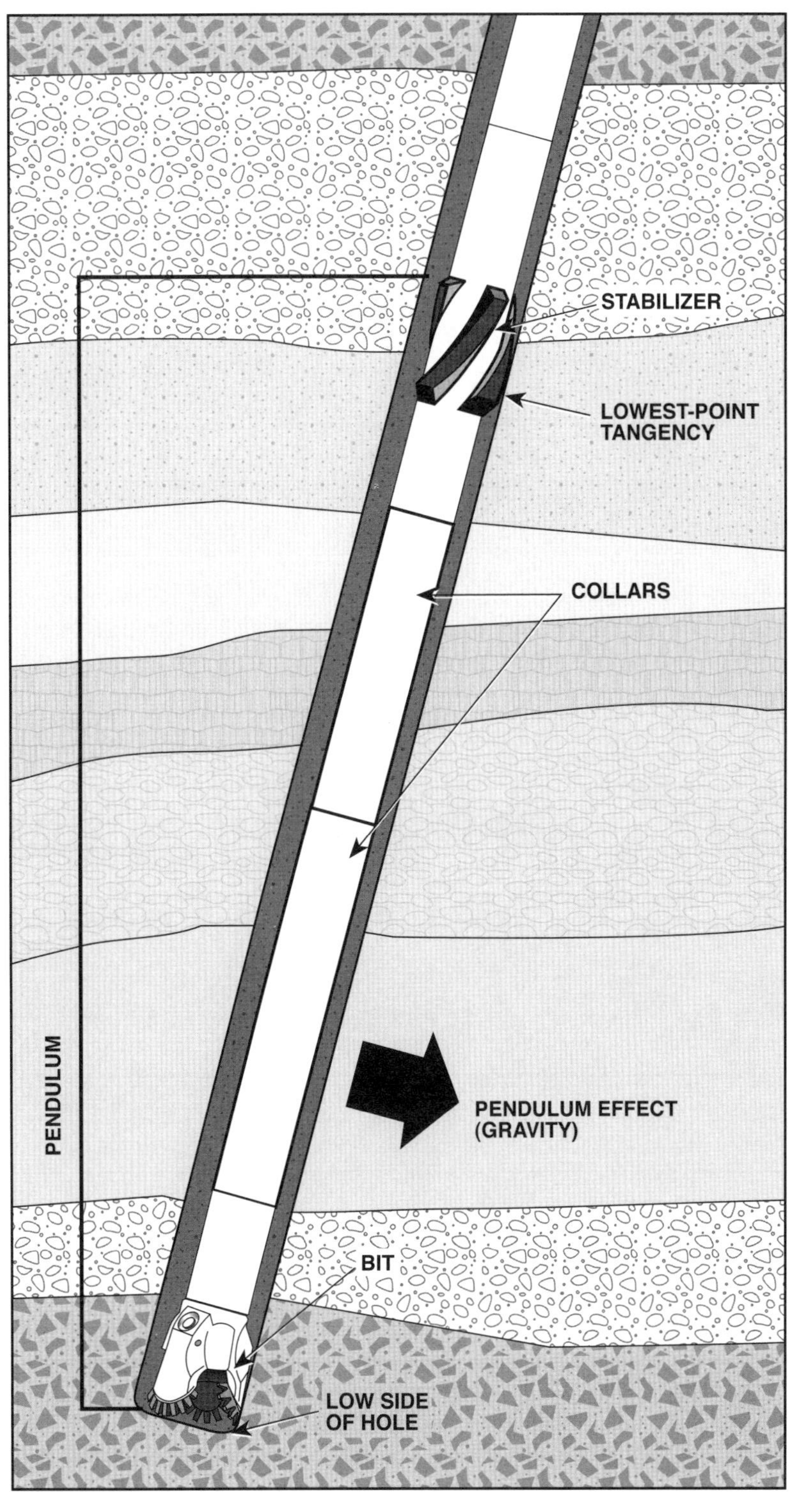

Figure 34. The first stabilizer acts as the point of tangency (fulcrum) below which gravity pulls the bit to the low side of the hole, producing a pendulum effect.

Table 1
8¾" Hole Size—15° Formation Dip
Drill Collar Sizes

HOLE ANGLE AND	CLASS.	6"	6" with Stabilizer at Height Shown		6½"	6½" with Stabilizer at Height Shown		7"	7" with Stabilizer at Height Shown		7½"	7½" with Stabilizer at Height Shown		8"	8" with Stabilizer at Height Shown	
2°	A	—	—	—	—	—	—	—	—	—	—	—	—	—	—	—
	B	—	—	—	—	—	—	—	—	—	—	—	—	—	—	—
	C	—	—	—	—	—	—	—	—	—	—	—	—	—	—	—
	D	—	—	—	—	—	—	—	—	—	—	—	—	—	—	—
	E	—	—	—	—	—	—	—	—	—	1,076	1,332	85–94	1,120	1,400	76–85
	F	—	—	—	—	—	—	—	—	—	1,345	1,668	85–94	1,398	1,746	76–85
	G	—	—	—	—	—	—	—	—	—	1,726	2,140	85–94	1,795	2,245	76–85
	H	—	—	—	—	—	—	—	—	—	2,219	2,746	85–94	2,305	2,880	76–85
	I	—	—	—	—	—	—	—	—	—	2,835	3,513	84–93	2945	3,680	76–85
	J	—	—	—	—	—	—	—	—	—	3,584	4,440	84–93	3,729	4,660	76–85
	K	3,027	3,720	90–100	3,547	4,390	90–100	4,050	4,980	87–97	4,476	5,550	84–93	4,684	5,850	76–84
	L	3,869	4,810	89–99	4,556	5,660	88–98	5,233	6,470	86–96	5,823	7,280	83–92	6,142	7,710	76–84
	M	4,925	6,020	88–98	5,826	7,280	86–96	6,727	8,360	85–95	7,519	9,410	82–91	7,918	9,960	75–83
	N	6,137	7,700	86–96	7,313	9,210	85–95	8,517	10,640	85–94	9,610	12,090	81–90	10,203	12,980	75–83
	O	7,679	9,760	84–93	9,223	11,700	84–93	10,845	13,670	84–93	12,381	15,700	80–89	13,349	16,920	74–82
	P	9,711	12,600	81–90	11,803	15,230	81–90	14,074	17,930	81–90	16,327	20,900	78–87	17,912	22,900	73–81
	Q	11,813	15,600	78–87	14,550	19,200	78–87	17,625	23,040	79–88	20,850	27,100	76–85	23,473	30,360	71–79
	R	15,031	21,200	71–79	18,891	26,200	73–81	23,434	31,800	75–83	28,453	38,200	73–81	32,768	43,200	68–76
	S	18,277	28,580	60–67	23,542	35,000	64–71	30,066	44,300	67–74	37,821	52,900	69–77	45,579	62,000	67–74
	T	22,433	40,300	49–55	29,863	53,750	52–58	39,654	70,500	52–58	52,598	85,000	55–61	69,306	103,200	56–62
	U	28,338	50,900	44–49	39,596	71,200	41–46	55,684	99,000	44–49	80,269	144,500	40–45	123,768	—	—
3°	A	—	—	—	—	—	—	—	—	—	—	—	—	—	—	—
	B	—	—	—	—	—	—	—	—	—	—	—	—	—	—	—
	C	—	—	—	—	—	—	904	1,124	80–89	991	1,232	76–85	1,031	1,290	69–77
	D	—	—	—	—	—	—	1,119	1,390	80–89	1,227	1,529	76–85	1,277	1,594	69–77
	E	—	—	—	—	—	—	1,412	1,755	80–89	1,548	1,930	76–85	1,611	2,017	69–77
	F	—	—	—	—	—	—	1,811	2,250	80–89	1,985	2,474	76–85	2,063	2,580	68–76
	G	—	—	—	—	—	—	2,304	2,860	80–89	2,525	3,150	76–85	2,625	3,280	68–76
	H	—	—	—	—	—	—	2,933	3,640	79–88	3,217	4,000	76–84	3,346	4,180	68–76
	I	2,756	3,390	81–90	3,230	4,010	81–90	3,681	4,570	79–88	4,054	5,050	76–84	4,236	5,290	68–76
	J	3,470	4,310	81–90	4,084	5,090	79–88	4,676	5,820	79–88	5,174	6,450	75–83	5,436	6,780	68–76
	K	4,346	5,550	80–89	5,137	6,420	78–87	5,899	7,340	78–87	6,529	8,150	75–83	6,865	8,570	68–76
	L	5,470	6,950	79–88	6,506	8,170	78–87	7,512	9,400	77–86	8,342	10,490	75–83	8,817	11,020	67–75
	M	6,781	8,550	77–86	8,122	10,300	77–86	9,451	11,880	76–85	10,592	13,330	74–82	11,316	14,200	67–74
	N	8,438	10,810	76–84	10,201	13,100	76–85	11,974	15,100	75–83	13,520	17,100	73–81	14,595	18,430	67–74
	O	10,386	13,420	74–82	12,695	16,440	75–83	15,092	19,300	74–82	17,308	22,200	71–79	19,053	24,300	66–73
	P	12,922	17,020	72–80	16,014	20,900	73–81	19,325	25,250	71–79	22,539	29,160	69–77	25,363	32,800	65–72
	Q	15,976	21,940	69–77	20,163	27,500	69–77	24,736	33,060	69–77	29,228	38,200	67–75	33,418	43,400	63–70
	R	19,552	28,800	61–68	25,194	36,660	63–70	31,640	44,300	66–73	38,436	52,200	65–72	45,612	60,600	61–68
	S	23,410	39,700	50–56	30,794	48,900	54–60	39,745	60,000	57–63	50,225	71,800	59–66	62,920	87,300	58–64
	T	28,144	50,600	44–49	38,139	68,500	43–48	51,027	91,800	46–51	67,581	113,400	48–53	90,811	136,000	45–50
	U	34,376	61,800	39–43	48,033	86,300	33–37	67,499	121,300	40–45	97,152	—	—	149,358	—	—
4°	A	—	—	—	655	816	76–85	743	924	75–83	815	1,016	71–79	847	1,060	65–72
	B	—	—	—	835	1,040	76–84	946	1,174	75–83	1,038	1,293	71–79	1,080	1,350	65–72
	C	—	—	—	1,043	1,302	76–84	1,182	1,472	75–83	1,296	1,615	71–79	1,351	1,688	65–72
	D	—	—	—	1,347	1,680	76–84	1,527	1,900	75–83	1,674	2,080	71–79	1,741	2,172	65–72
	E	—	—	—	1,671	2,080	76–84	1,892	2,357	75–83	2,073	2,572	70–78	2,158	2,695	64–71
	F	—	—	—	2,137	2,660	76–84	2,421	3,013	74–82	2,654	3,296	70–78	2,762	3,450	64–71
	G	2,303	2,860	76–84	2,691	3,350	75–83	3,055	3,800	74–82	3,356	4,170	70–78	3,489	4,360	64–71
	H	2,891	3,580	76–84	3,387	4,210	75–83	3,860	4,810	74–82	4,261	5,310	70–78	4,448	5,550	64–71
	I	3,682	4,580	75–83	4,333	5,410	75–83	4,952	6,160	73–81	5,487	6,830	70–78	5,730	7,150	64–71
	J	4,613	5,770	75–83	5,437	6,780	74–82	6,207	7,760	73–81	6,868	8,580	70–78	7,207	9,000	64–71
	K	5,653	7,100	73–81	6,699	8,440	74–82	7,684	9,630	72–80	8,554	10,720	69–77	9,022	11,290	63–70
	L	7,066	8,950	72–80	8,431	10,620	73–81	9,747	12,300	71–79	10,959	13,800	68–76	11,631	14,640	63–70
	M	8,807	11,280	71–79	10,591	13,400	72–80	12,330	15,670	70–78	13,985	17,650	67–75	14,994	19,000	62–69
	N	10,725	13,810	69–77	13,031	16,840	70–78	15,348	19,640	69–77	17,668	22,500	67–74	19,244	24,600	61–68
	O	13,237	17,300	68–76	16,262	21,100	68–76	19,374	25,000	67–75	22,630	29,100	66–73	24,987	32,000	61–68
	P	16,680	22,360	67–74	20,790	27,600	66–73	25,042	33,100	66–73	29,661	38,800	64–71	33,118	42,900	60–67
	Q	20,022	28,400	60–67	25,357	35,400	63–70	31,087	42,500	63–70	37,657	50,500	61–68	43,132	56,800	58–65
	R	24,000	37,200	52–58	31,035	46,200	56–62	39,162	55,900	59–66	49,214	68,500	58–65	58,882	80,500	56–62
	S	28,203	50,700	43–48	37,256	59,900	48–53	48,304	75,300	50–56	[illegible]	[illegible]	[illegible]	[illegible]	[illegible]	[illegible]

Table 1 is for an 8¾-in. (222.25-mm) hole in a formation with 15-degree dip. It allows you to predict how much weight on bit you could run with various drill collar sizes and stabilizer placement and continue to maintain hole angle. Be aware the tables are based on established equilibrium drilling conditions, as in the earlier example, and are valid for only those conditions. Also, if you place a reamer or stabilizer near the bit, it negates the predicted effects.

To use the table, find the hole inclination angle of 4 degrees and collar size of 6 in. (152.4 mm). Go down the 6-in. (152.4-mm) column to 13,237 lb (5,850.75 dN), the weight on bit. This is line, or class, O. Any other point on line O on this table is comparable to the current condition. Thus, if you changed to 7½-in. (190.5-mm) drill collars, you could carry 22,630 lb (10,002.5 dN) weight on bit and still maintain 4 degrees of hole inclination. If you placed a stabilizer 66 ft (20.1 m) above the bit, you could carry 29,100 lb (12,862.2 dN) weight on bit. The decision to change to larger collars and stabilizers must always be an economic one. Will the faster penetration rate justify the extra expense?

Line O on the 3-degree table shows that you could reduce weight on bit to 10,368 lb (4,582.7 dN) and eventually drop the hole angle to 3 degrees with the same 6-in. (152.4-mm) collars.

Although a single stabilizer properly placed in the string helps control deviation, operators usually place a second or third stabilizer in the drill collar string at 30 ft (9 m) and 60 ft (18 m) above the fulcrum stabilizer. Adding additional stabilizers is usually desirable to reduce lateral force on the fulcrum stabilizer. The additional stabilizers make the fulcrum stabilizer more effective by preventing it from digging into the wall of the hole.

Operators generally use the pendulum assembly for drilling soft, unconsolidated formations where fast penetration rates can be maintained with lighter weights on bit. They also use it as a corrective measure to reduce angle when deviation exceeds the maximum allowed. Directional drilling service companies offer tables and computer programs that predict the hole angle drop rate for various conditions (table 2).

A pendulum assembly by itself does not automatically prevent a dogleg from occurring. Only when formation force and pendulum force are equal will the bit drill straight ahead in the same direction. However, even under equilibrium, the bit can move freely from side to side in a washed-out soft formation. This lateral movement continues until the drill collars come into contact with hard rocks in the wall of the hole above to stop it (see fig. 20).

Table 2
Angle-Drop Rates to Correct Hole Deviation

ANGLE DROP 90°	9.75-IN. HOLE WITH 7.25-IN. × 2.25-IN. DRILL COLLARS	
Existing Hole Inclination Angle	**Weight on Bit**	**Estimated Drop Rate/100 ft**
30°–45°	0–15,000 lbs 15,000–30,000 lbs	2.00°–2.50° 1.25°–1.50°
20°–30°	0–15,000 lbs 15,000–30,000 lbs	1.25°–1.50° 0.75°–1.00°
5°–20°	0–15,000 lbs 15,000–30,000 lbs	0.75°–1.00° 0.50°–0.75°
0°–5°	0–15,000 lbs 15,000–30,000 lbs	0.00°–0.50° 0.00°–0.00°

ANGLE DROP 60°	9.75-IN. HOLE WITH 7.25-IN. × 2.75-IN. DRILL COLLARS	
Existing Hole Inclination Angle	**Weight on Bit**	**Estimated Drop Rate/100 ft**
30°–45°	0–15,000 lbs 15,000–30,000 lbs	1.25° 1.00°
20°–30°	0–15,000 lbs 15,000–30,000 lbs	1.00° 0.75°
0°–20°	0–15,000 lbs 15,000–30,000 lbs	0.75° 0.50°

ANGLE DROP 30°	9.75-IN. HOLE WITH 7.25-IN. × 2.75-IN. DRILL COLLARS (with undergauge near-bit stabilizer)	
Existing Hole Inclination Angle	**Weight on Bit**	**Estimated Drop Rate/100 ft**
20°–45°	0–15,000 lbs 15,000–30,000 lbs	0.75° 0.50°
0°–20°	0–15,000 lbs 15,000–30,000 lbs	0.25° 0.25°

Courtesy of Sperry-Sun Drilling Services, a Halliburton Company

Packed-Hole Assembly

Today, operators usually drill wells in crooked-hole country with some type of packed-hole assembly (PHA) because it permits maximum weight to be run on the bit for fast rates of penetration. Keep in mind, however, that even the smallest amount of bending causes a bit to deviate and PHAs bend—not very much perhaps, but they do bend. Therefore, even with a PHA, a perfectly vertical hole is not possible. Nevertheless, a properly designed PHA keeps the rate of hole angle change to a minimum, and prevents doglegs.

Packed means that the drill collars or stabilizers in the lower part of the assembly are no more than ⅛ in. (3 mm) smaller in diameter than the hole. Operators sometimes refer to using a PHA as the gun barrel approach, because a PHA keeps the hole "as straight as a gun barrel."

A PHA run above the bit must have the necessary stiffness and wall contact to force the bit to drill in the general direction the hole is already heading. If operators select the proper drill collars and bottomhole tools, only gradual changes in hole angle will develop and doglegs will be avoided.

A properly designed PHA—

1. reduces the rate of hole angle change;
2. improves the performance and life of the bit by forcing it to rotate on a true axis about its design center, thus loading all cones equally;
3. improves hole conditions for drilling, logging, and running casing; and
4. allows the use of more drilling weight in formations that cause abnormal drift.

Design Factors

Operators usually run a PHA into the hole to ensure that further deepening keeps it on course with the desired amount of inclination. If operators wish to maintain the angle and drift of the hole, they must pay particular attention to several factors when designing a PHA. Factors include (1) the length of contact of the assembly, (2) the stiffness of the drill collar string, (3) the clearance between the hole and the stabilizers, and (4) the availability of wall support for the stabilizers.

PHAs must provide sufficient length of contact with the wall of the hole to assure alignment with the hole already drilled. If an operator used only a single stabilizer just above the bit, hole angle would build. Hole angle would build because the lateral force of the unstabilized collars above the bit causes the bit to push to one side as weight is applied. Placing another stabilizer at 30 ft (9 m) above the bit nullifies some of the angle-building tendency and stabilizes the bit; however, this assembly still allows the hole to deviate. Only an assembly with three stabilization points can ensure alignment with a previously drilled hole (fig. 35).

Drill collar stiffness is a second important design consideration. Stiff drill collars prevent doglegs by adding stiffness to the BHA and by stabilizing the bit. The largest diameter collars that can safely be run should be selected, since drill collars of larger

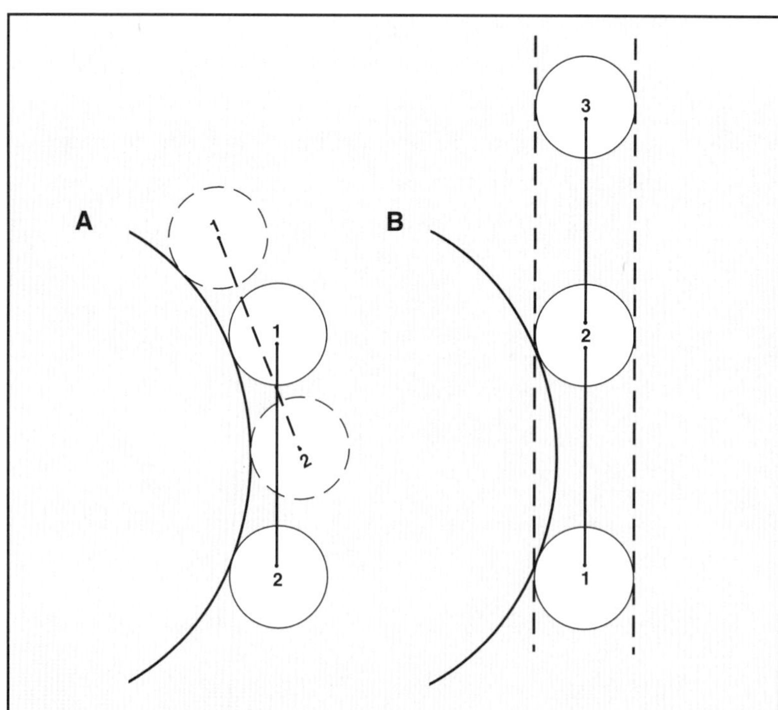

Figure 35. Stabilization points. Two points can contact and follow a curved line (A), but the addition of one more point eliminates this tendency (B).

Table 3
Moments of inertia for popular-sized drill collars

Collar OD (in.)	Collar ID (in.)	Moments of Inertia
5	2¼	29
6¼	2¼	74
6½	2¼	86
6¾	2¼	100
7	2¹³⁄16	115
8	2 13/16	198
9	2 13/16	318
10	3	486
11	3	713

diameters provide the most stiffness. Square collars provide more stiffness than round collars of equal size.

Standard drill collars increase in stiffness by the fourth power of their diameter. Table 3 lists the moments of inertia, which are proportional to stiffness, for popular collar sizes. You can see from Table 3 that a drill collar with an outside diameter (OD) of 9 in. (228.6 mm) is nearly three times stiffer than a 7-in. (177.8-mm) OD drill collar with the same inside diameter (ID). Keep in mind, however, that even the stiffest drill collar will bend (fig. 36).

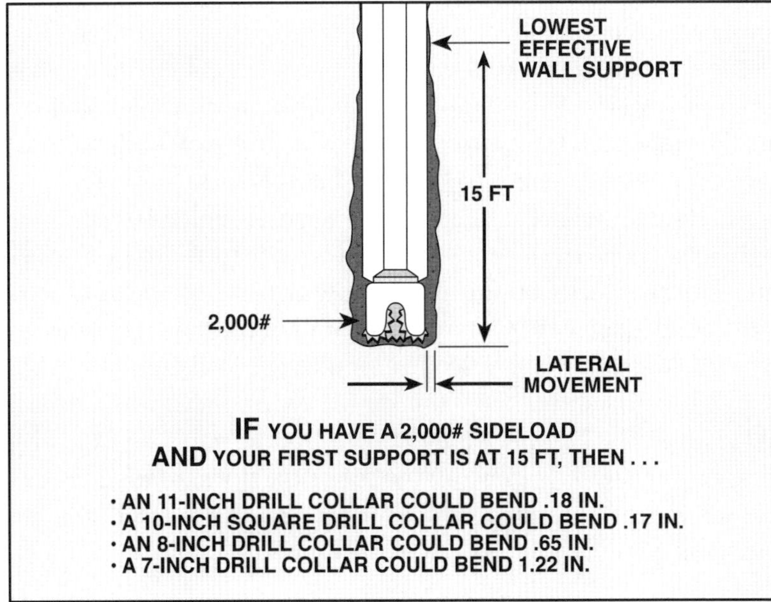

Figure 36. Relative deflection of various size drill collars. In each case the drill collar is supported 15 in. (381 millimetres) above the bit and subjected to a 2,000-lb (900-kilogram) sideload. (Courtesy of Smith International, Inc.)

Also keep in mind that certain factors limit the maximum size of drill collars that the operator can run. For example, small rigs cannot handle large collars without special rental equipment, which requires extra time and expense. Further, operators must size drill collars so that sufficient clearance between the drill collar and the wall of the hole is available for fishing with an overshot or a washpipe. Enough room must exist to allow an overshot or a washpipe to fit over the drill collar. Another problem with oversize collars is increased hole erosion in some formations. The restricted annular clearance between the collar and the wall of the hole can produce higher fluid velocities that cause excessive washout.

The change from large drill collars to smaller collars, or from collars to drill pipe, in the upper end of the drill collar string should be gradual. A gradual transition zone can avoid the problems of connection failure or drill string fatigue (see fig. 46) caused by too sudden a change in the size of the components in the drill string. Rigid, heavy collars do not bend as much as the more limber drill pipe string. Most of the bending stress is on the first few joints of drill pipe immediately above the drill collars. To minimize bending stresses on the drill pipe in the transition zone, contractors often use heavy-walled drill pipe or heavyweight drill pipe between the drill collars and the conventional drill pipe. Heavy-walled drill pipe has much thicker walls than conventional pipe and can better withstand bending stresses than conventional pipe. It also absorbs bending stresses before they reach the conventional pipe.

Another important design consideration is the amount of clearance between the stabilizers and the wall of the hole. The clearance must be minimal in order for the stabilizers to work effectively. The closer the stabilizer is to the bit, the more exacting the clearance requirements are. For example, if ⅛-in. (3-mm) clearance 60 ft (18 m) above the bit is satisfactory, then 1⁄16-in. (1.5-mm) clearance just above the bit is required to keep the bit on course.

A fourth design consideration for a packed-hole assembly is the wall support that the formation provides for the stabilizers. Formations offering little support may simply crumble away. Stabilizers must be adequately supported by the wall of the hole if they are to effectively stabilize the bit and centralize the drill collars.

The surface area of a stabilizer that is in contact with the wall of the hole must be large enough to prevent the stabilizer from digging into the wall. In general, the softer the formation, the larger the surface area required. If the surface area is not large enough, the stabilizer digs into the wall of the hole and stabilization is lost. As a result, the hole drifts off course.

If a formation is hard and uniform, it provides strong wall support for a stabilizer. A short and narrow surface on the stabilizer is adequate to ensure proper stabilization. On the other hand, a soft and unconsolidated formation provides poor wall support, and a long-blade stabilizer with a large surface area is required.

Soft formations may present an additional problem. If they erode quickly, a hole drilled through them may enlarge. Hole enlargement reduces effective alignment of the BHA. Drill collars that are too close to bit size cause higher annular fluid velocities and greater hole erosion. Square or spiral drill collars and controlling drilling fluid properties can alleviate the problem.

Packed-hole Assemblies for Various Conditions

Operators make up PHAs with combinations of stabilizers, depending on the crooked-hole tendencies of the formations to be penetrated and the drillability of the formations. Formations with mild, medium, or severe crooked-hole tendencies require different PHAs (fig. 37).

A PHA usually consists of a bit, a short drill collar, a normal-length drill collar, and several stabilizers placed at various points in the drill collar string. When referring to placement of stabilizers in a PHA, operators often divide the PHA into three zones. Zone 1 is near the bit, zone 2 is above a short drill collar installed above the bit, and zone 3 is above a drill collar(s) of normal length, which is installed above the short drill collar or a vibration dampener if used. Thus, crew members can make up a PHA with three or more large-diameter stabilizers and a short drill collar between the near-bit stabilizers in zone 1 and the stabilizer in zone 2.

When using the English system of measurement, a rule for determining the length of the short drill collar is that it should be the same number of feet as the number of inches of the hole diameter, plus or minus 2 ft. Therefore, a short collar length of 6 to 10 ft would be satisfactory in an 8-in. hole.

The number of stabilizers used and their placement in the BHA depends on the severity of the crooked-hole tendencies of the formation. As crooked-hole tendencies become more severe, additional stabilizers are required in zone 1 to prevent bit deviation. The additional stabilizers increase the area in contact with the wall of the hole; using stabilizers with longer and wider blades may further increase wall contact area. Any stabilizers run above zone 3 are used only to prevent drill collars from buckling or becoming stuck. They have very little effect on directing the bit.

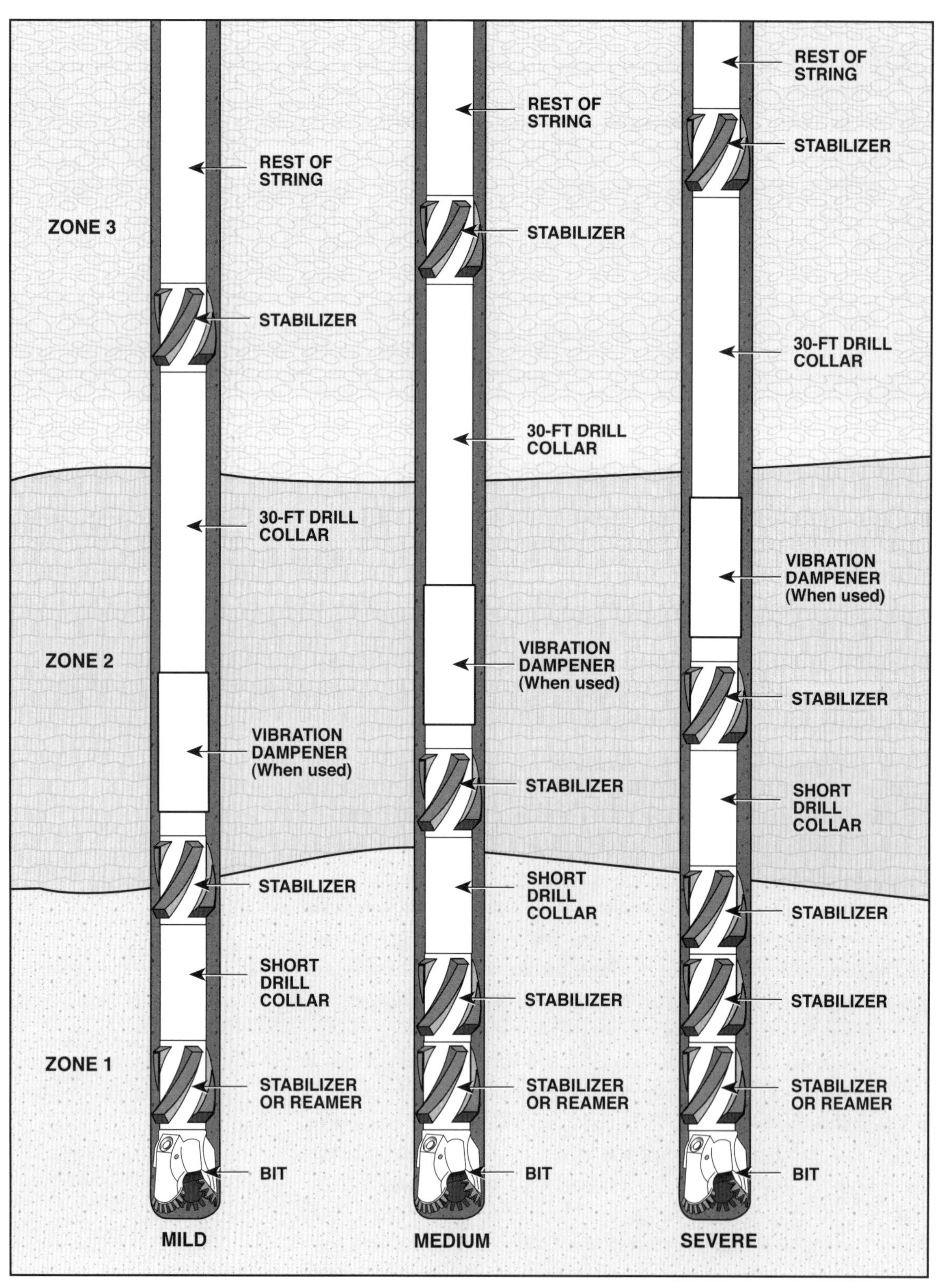

Figure 37. Packed-hole assemblies for mild, medium, and severe crooked-hole conditions

Hard, abrasive formations wear down the sides of the bit. The bit then drills an undergauge hole where full-gauge bits and tools can get stuck. To minimize this problem, operators usually install reamers or reamer-stabilizers in the BHA to ensure that the hole is drilled full size (fig. 38).

For straight-hole drilling and bit stabilization in mild crooked-hole country, a minimal PHA is usually all that is required. In general, operators use three points of stabilization: one in zone 1 immediately above the bit; a second in zone 2 above the large-diameter, short drill collar; and a third in zone 3 above the large-diameter, regular-length drill collar.

Figure 38. A reamer-stabilizer is run to minimize the drilling of undergauge hole.

When the operator elects to run a vibration dampener, crew members should place it above zone 2 where it will work most effectively by reducing the shock loads on the bit cutters and drill collar connections. If crooked-hole tendencies are mild, the vibration dampener may be run in place of the short drill collar between zones 1 and 2. Vibration dampeners increase penetration rates and add life to the bit. They also reduce wear and damage to the drilling rig and drill string.

A PHA for medium crooked-hole country requires the addition of a second stabilizer in zone 1. The two tools run together provide increased bit stabilization and add stiffness to limit angle changes caused by lateral forces.

In severe crooked-hole country, three stabilization tools are run in zone 1, which provide maximum stiffness and wall contact area. In hole sizes of 8¾ in. (222.25 mm) and smaller, most operators run a large-diameter, short drill collar between zones 2 and 3. The collar increases stiffness and reduces deflection of the total assembly; it also allows tools in zones 1 and 2 to function without excessive wear caused by lateral thrust or side-loading from excess deflection above.

In areas with steep dip, a PHA is needed to stay within contract deviation limits. Figures 39 and 40 show predicted performance in formations dipping at 20 degrees and at 40 degrees. In the 20-degree case, note that the slick assembly will reach the 7-degree maximum limit far above the expected 7,000-ft (2,134-m) total depth (TD). The PHA should reach TD within the deviation limit.

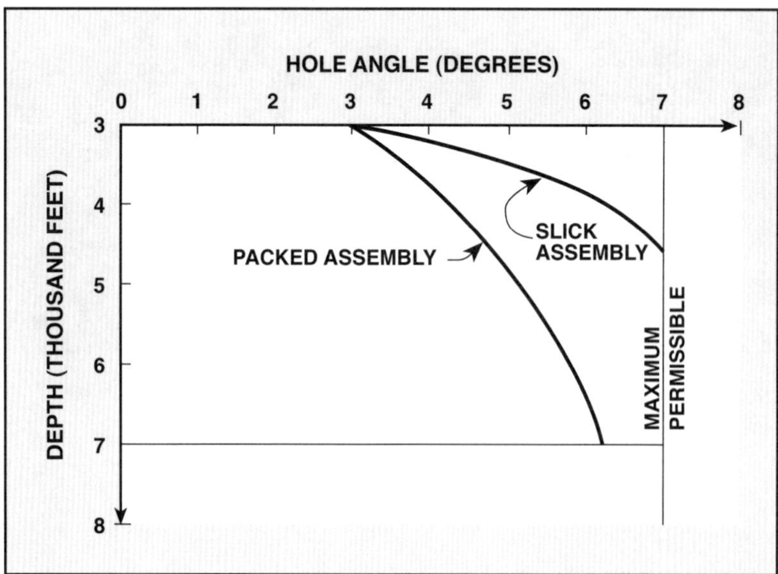

Figure 39. Predicted performance of BHAs in case study with 20-degree formation dip (Courtesy of Society of Petroleum Engineers)

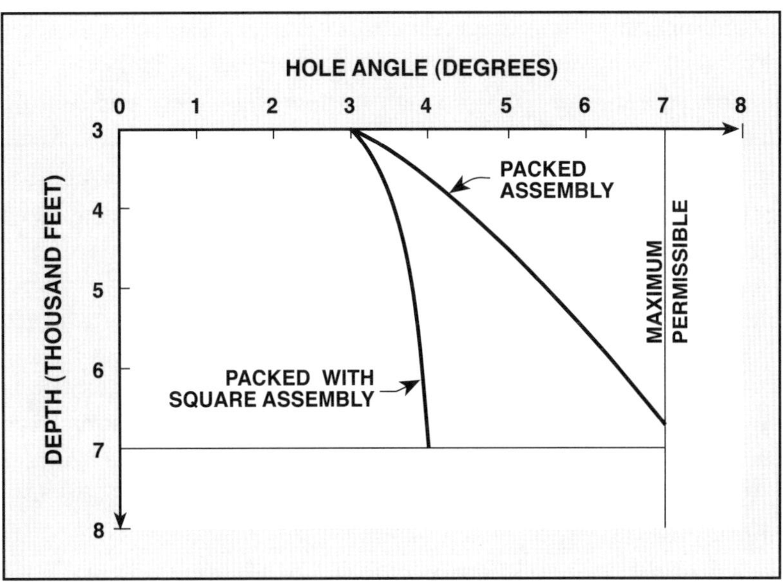

Figure 40. Predicted performance of BHAs in case study with 40-degree formation dip (Courtesy of Society of Petroleum Engineers)

In the 40-degree dip case, a PHA will exceed the deviation limit before reaching TD. The inclusion of square drill collars in the PHA greatly reduced expected hole deviation; it had about 4 degrees deviation at TD (fig. 41).

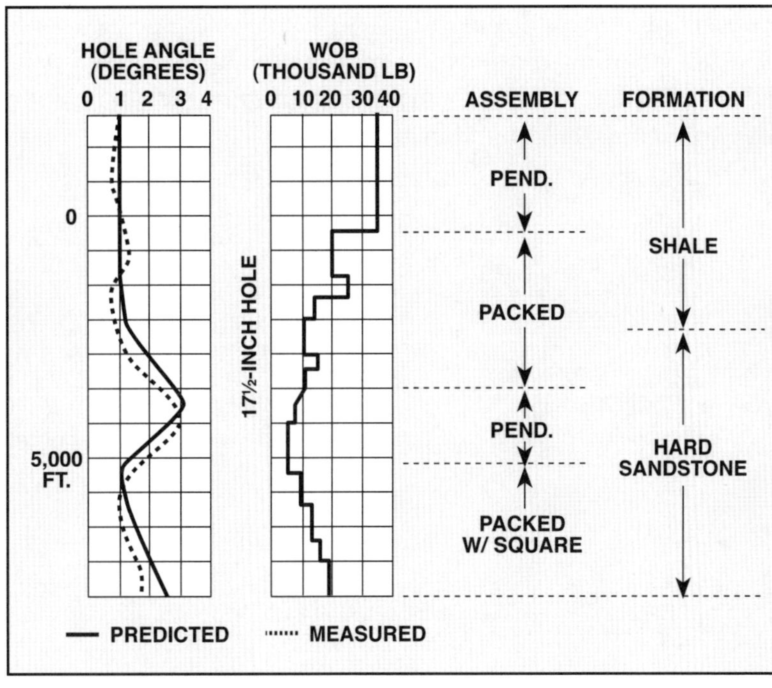

Figure 41. A straight hole field case with 4-degree maximum inclination limit, southern Oklahoma. Illustration shows predicted and measured-hole inclination effects of various BHAs, formations, and WOB combinations. (Courtesy of Society of Petroleum Engineers)

Sometimes it is necessary to change BHAs several times during the course of drilling, as in the southern Oklahoma case illustrated in figure 10. Note that the hole angle increased sharply at 4,800 ft (1,463 m) in the hard sandstone, approaching the 4-degree maximum allowed in the contract. A pendulum assembly quickly brought the hole back close to vertical and the packed BHA included square drill collars to TD.

Packed Pendulum Assembly

PHAs keep the rate of hole angle change to a minimum. Occasionally, however, the operator must reduce total hole deviation. They can use the pendulum technique to reduce total hole angle, and, if a PHA is required after the pendulum technique reduces the hole angle, they can use a packed pendulum assembly.

In a packed pendulum assembly, a pendulum of from one to three drill collars is swung below the regular PHA (fig. 42). The angle drop (correction toward vertical) depends on placement of the first stabilizer (which creates a fulcrum point) above the bit, existing inclination conditions, and weight on bit. Table 2 illustrates the angle drop for three pendulum lengths under different inclination angles and weights on the bit. When hole deviation has dropped to an acceptable limit, crew members remove the pendulum collars, and again run a PHA above the bit. With this technique, the operator only has to ream the length of the pendulum collars prior to resuming normal drilling.

If a vibration-dampening device is used in the packed pendulum assembly, it should remain in its original pickup position when resuming the pendulum operations.

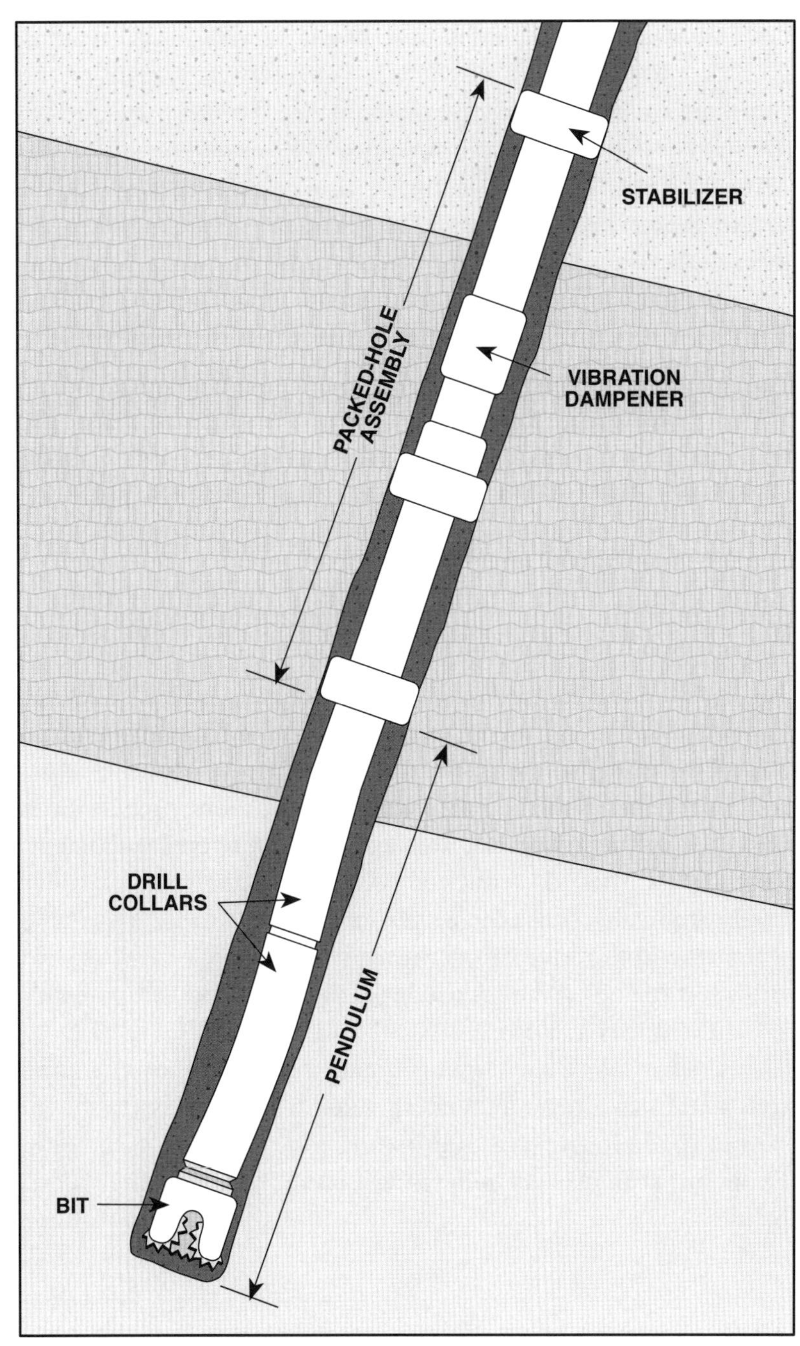

Figure 42. A packed pendulum assembly

Downhole Mud-Motor Assemblies

A downhole motor is a length of strong pipe with an OD that resembles a drill collar. Inside the pipe is a helical device called the rotor-stator or, in turbines, a series of blades. The bit is attached to a drive shaft that is connected directly to the motor. Pump pressure forces mud past the rotor-stator, or through the turbine blades, turning them along with the bit. The bit rotates but the drill string does not. Downhole motors require more pump capacity than regular rotary drilling because the bit is normally rotated at extremely high speeds (up to 2,000 revolutions per minute or rpm).

The drilling assembly usually includes a bent housing or an adjustable kickoff (AKO) tool. The AKO is placed between the downhole motor and the bit (fig. 43). Using information from previous borehole surveys or MWD data, the correct angle and direction are set up in the bent housing. Drilling is resumed and monitored via MWD telemetry.

Reduced Bit Weight

One of the oldest techniques for straightening a hole is to reduce weight on the bit and speed up the rotary table. When weight on the bit is reduced, bending characteristics of the drill string change, and the hole tends to be straighter. However, reducing bit weight also reduces the penetration rate and frequently creates doglegs, especially if the weight reduction is sudden (see fig. 7). The driller should gradually reduce the weight on the bit to straighten a hole so that the hole returns to vertical without sharp bends. A hole with a severe dogleg can make completing and producing the well very difficult (see tables 1 and 2).

A reduction of bit weight is usually required when changing from a PHA to a pendulum or a packed pendulum assembly. To prevent the hole angle from dropping too quickly in such cases, the operator can place an undergauge stabilizer immediately above the bit.

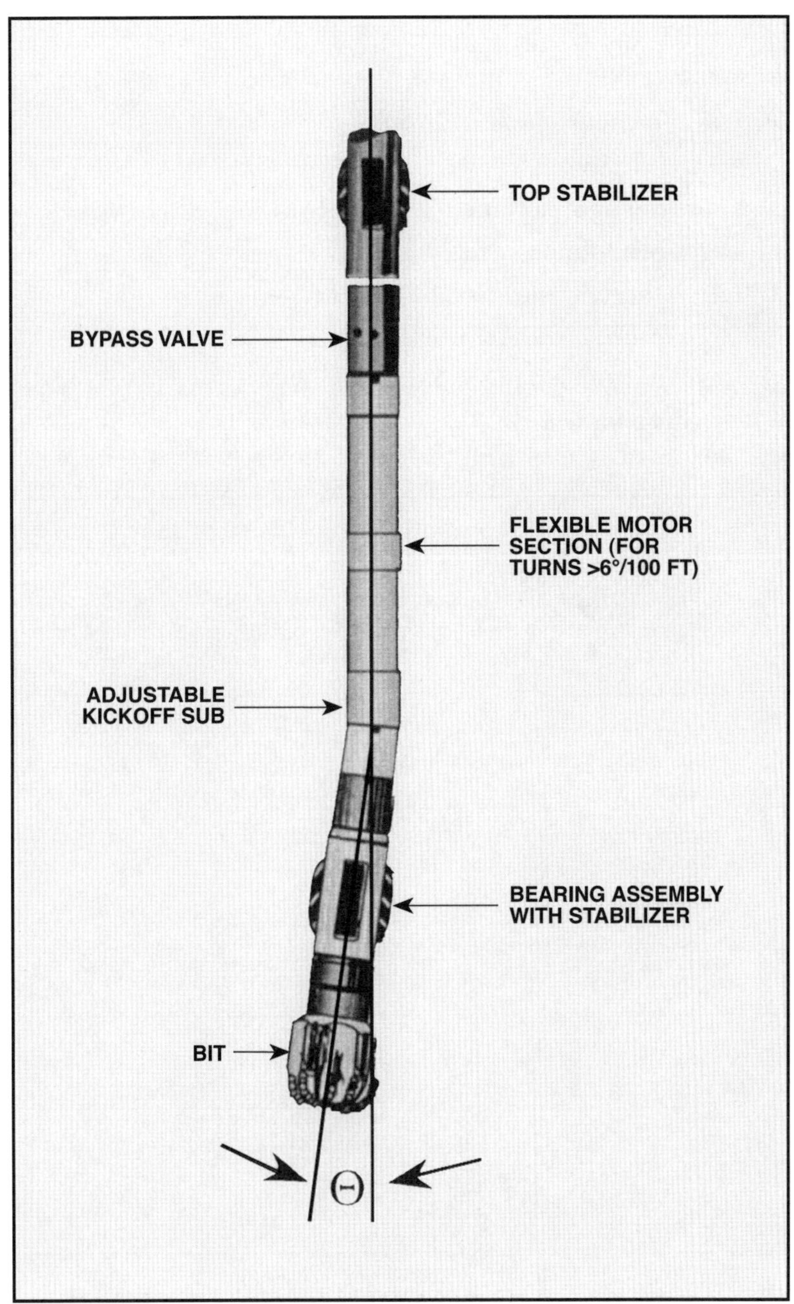

Figure 43. An adjustable kickoff (AKO) downhole motor assembly can be used for drilling slant and horizontal holes and for correcting excessive deviation in vertical holes. (Courtesy of Baker Hughes INTEQ)

To summarize—

Methods of controlling hole deviation

- start straight
- use the right bottomhole assembly (BHA); BHAs include—
    1. pendulum assembly
    2. packed-hole assembly (PHA)
    3. packed pendulum assembly
    4. downhole mud-motor assemblies
- reduce bit weight

Drill Stem Tools

Operators use several drill stem tools to control deviation and drill straight holes. Such tools include standard drill collars, heavy-walled and heavyweight drill pipe, square drill collars, spiral drill collars, stabilizers, vibration dampeners, and measurement while drilling tools.

Standard Drill Collars

Today, operators use long strings of drill collars virtually everywhere, even in soft formations. Drill collars supply weight to the bit for drilling and maintain weight to keep the drill string from bending or buckling (fig. 44). They also prevent doglegs by adding stiffness to the BHA and by supporting and stabilizing the bit.

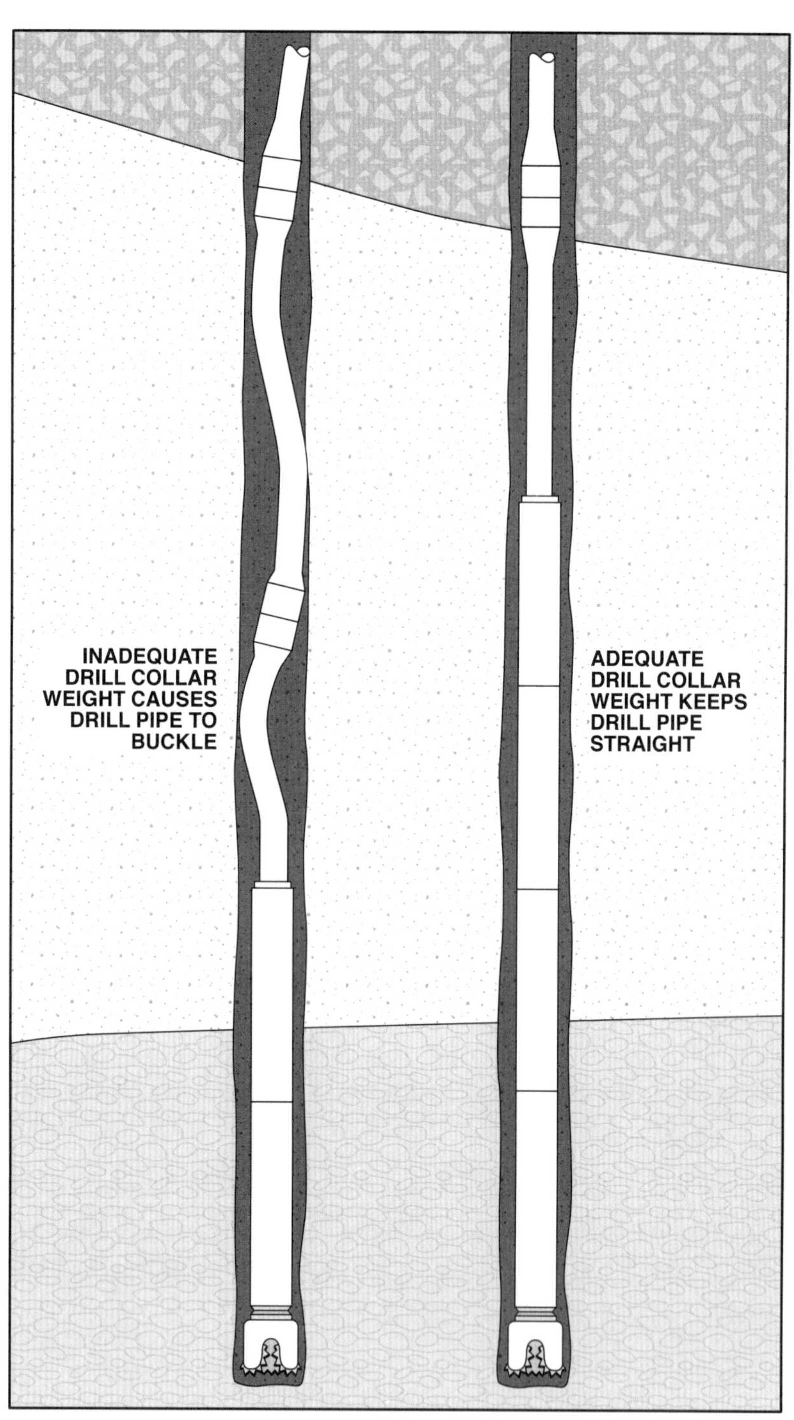

Figure 44. Effects of drill collar weight

Drill Collar Size and Weight

Most collars are either 30 or 31 ft (9.14 or 9.45 m) long; their weight varies considerably and depends on the OD and ID (bore) of the drill collar. For example, one of the lightest collars has an OD of 3 in. (76.2 mm) and weighs 640 pounds (lb) or 290.3 kilograms (kg). One of the heaviest has a 12-in. (304.8-mm) OD and weighs 11,400 lb (5,171 kg). Drill collar weight ranges from 21 to 368 pounds per foot (lb/ft) or 31.25 to 547.58 kilograms per metre (kg/m) (table 4). The ID, or bore, of a drill collar varies from 1 to 3¼ in. (25.4 to 82.6 mm). Also, a small, thin-wall drill collar may have the same ID as a larger thick-wall collar (fig. 45).

Operators must select drill collars with IDs that are large enough to handle the required circulation rates without excessive pressure drop. Collars with small IDs create more pressure drop than collars with large IDs. Too much pressure drop in the drill collar string means less pressure at the bit to assist in cleaning the bottom of the hole. Drill collar bore must also be large enough to allow tools to pass through them. Large bore collars are preferred because, in general, small-diameter free-point indicators, string shots, and survey instruments are not as reliable as instruments of larger diameters. Moreover, larger bores are required to obtain accurate horizontal angle measurements; accurate vertical angle measurements can, however, be obtained with small-bore drill collars. Further, larger bores may also be required if the housing on instruments used in hot holes must be insulated.

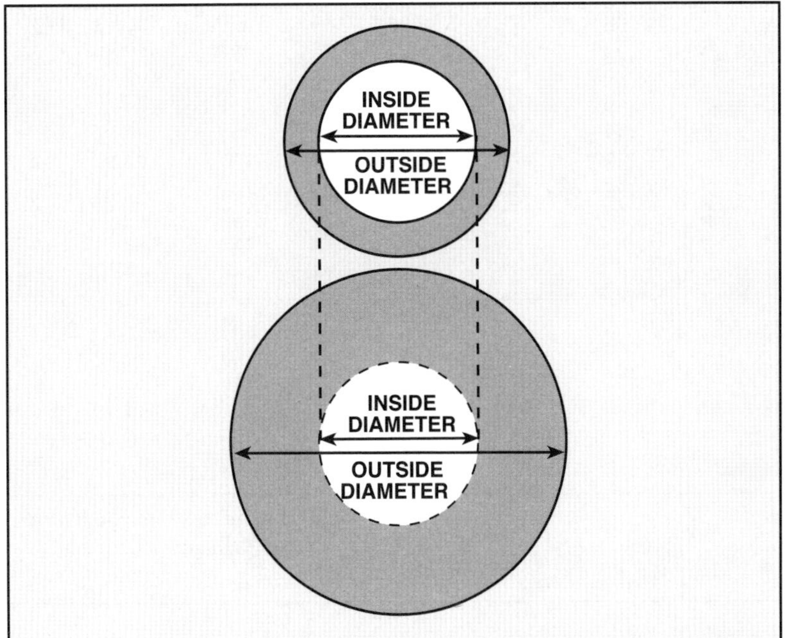

Figure 45. Two drill collars with the same inside diameter but different outside diameters

Table 4
Drill Collar Weights (lb/ft)

Drill Collar OD (in.)	Bore of Drill Collar (in.)										
	1	1⅛	1¼	1½	1¾	2	2¼	2½	2 13/16	3	3¼
3	21	21									
3⅛	23	23	22	21							
3¼	26	25	24	22							
3⅜			26	24	22						
3½			29	27	25						
3¾			33	32	29						
3⅞			36	34	32	30	27				
4				37	35	32	29				
4⅛				39	37	35	32				
4¼				42	40	38	35				
4½				48	46	43	41				
4¾				54	52	50	47	44			
5				61	59	56	53	50			
5¼				68	65	63	60	57	53		
5½				75	73	70	67	64	60	57	
5¾				83	80	78	75	72	67	64	
6				90	88	85	83	79	75	72	68
6¼				98	96	94	91	88	83	80	76
6½				107	105	102	99	96	92	88	85
6¾				116	114	111	108	105	101	98	94
7				125	123	120	117	114	110	107	103
7¼				134	132	130	127	124	119	116	112
7½				144	142	140	137	134	129	126	122
7¾				154	152	150	147	144	139	136	132
8				165	163	160	157	154	150	147	143
8¼				176	174	171	163	165	161	158	154
8½				187	185	182	179	176	172	169	165
8¾				193	196	194	191	188	183	180	176
9					208	206	203	200	195	192	188
9¼					220	218	215	212	207	204	200
9½					233	230	228	224	220	217	213
9¾					246	243	240	237	233	230	226
10						256	254	250	246	243	239
10¼						270	267	264	259	257	252
10½						284	281	278	273	270	266
10¾						298	295	292	287	285	280
11								306	302	299	295
11¼								321	317	314	310
11½								336	332	329	325
11¾								352	348	345	340
12								368	363	361	356

1,000 lb of steel displaces .364 barrel
65.5 lb of steel displaces one gallon
7.84 kg of steel displaces one litre

490 lb of steel displaces one cubic foot
2,747 lb of steel displaces one barrel

Courtesy of Drilco Grant Drilling Handbook

Drill collar OD ranges from 3 to 12 in. (76.2 to 304.8 mm). Drill collars of different ODs do not necessarily have different sized boxes or pins. The outside diameter of a collar should be as large as possible because a large-diameter collar provides more stiffness than a collar of a smaller diameter. Close fitting, stiff drill collars provide efficient bit performance; further, fatigue damage to the connections is less with collars that closely fit the hole. Collars with ODs close to hole size do not bend or flex as much as collars whose OD is considerably smaller than hole size.

Drill Collar Selection

To drill a straight hole, a drill collar string ideally would have the same OD as the bit size. However, this ideal is not practical because it would not allow drilling fluids to circulate up the hole. Still, operators should select drill collars with the largest outside diameter and the maximum permissible wall thickness that can safely be run in the hole and fished out in case of trouble.

In addition to providing more stiffness to the bottomhole assembly than small-diameter collars, large collars concentrate weight nearer to the bit where it is needed most. Long strings of small collars lose much of their apparent weight through friction and buckling in the hole. Only 14 collars with an OD of 10 in. (254 mm) and an ID of 3 in. (76.2 mm) are needed to supply as much weight as 25 collars with an OD of 7¾ in. (196.9 mm) and the same ID.

The threaded connections are the only part of standard drill collars that suffer rapid damage. Because fewer large-diameter drill collars are needed to supply an equal weight in small collars, fewer connections are made. If fourteen 10-in. (254-mm) OD collars are used instead of twenty-five 7¾-in. (196.9-mm) OD collars, the number of locations where stresses are high enough to cause fatigue failure is reduced by 44 percent. Fewer collars also require less time for handling during trips.

Although large collars are the best tools for combating crooked-hole problems, limitations may prevent the use of the very largest. For example, the contractor may have to rent special tools to pick up and lay down large collars. The time and expense spent handling the collars may be too great to justify the benefits gained by their use. An additional drawback to large drill collars is the possibility that they may cause increased hole erosion in some formations because of the high fluid velocity produced in the restricted annular area between the collars and the wall of the hole.

A slow penetration rate along with high pump pressure usually compounds the problem. Too large a collar may also restrict fluid and cuttings passage and cause the hole to pack off, a problem that may be prevented by use of square or spiral drill collars.

Large drill collars may also present a problem if they have to be fished out of the hole. Fishing can mean running an overshot, or it can mean washing over. The availability of washover pipe and overshot tools must be considered. Some sizes of washover pipe may not be available in the area where a well is being drilled (table 5).

Table 5
Maximum Drill Collar Size that Can Be Caught with Overshot and/or Washed Over with Washpipe

Hole Size, in.	Overshot		Washpipe		Maximum Fish OD to Catch or Washover, in.
	Size, in.	Max. Catch, in.	Size, in.	Max. Fish OD, in.	
6⅛	*5¾	5⅛	5½	4¾	4¾
6¼	*5¾	5⅛	5¾	4⅞	4⅞
6¾	*6⅜	5¼	6	5½	5½
7⅞	*7⅜	6¼	7	6⅛	6⅛
8⅜	*7⅞	6¾	7⅜	6½	6½
8½	*8	6⅞	7⅝	6¾	6¾
8¾	*8¼	7⅛	8⅛	7⅛	7⅛
9½	*9	7⅞	8⅝	7⅝	7⅝
9⅞	*9⅛	8	9	8	8
10⅝	*9¾	8⅝	9⅝	8½	8½
11	10½	8⅞	10¾	9⅝	9⅝
12¼	11¾	10⅛	11¾	10½	10⅛
13¾	12¾	11¼	12¾	11½	11¼
14¾	13¾	12	13⅜	12	12
17½	15⅛	13⅜	16	14½	13⅜
20	16¾	14¾	18⅝	17⅜	14¾
24	20¼	16¾	21	19½	16¾
26	24¾	22	21	19½	19½

*Overshots are not full strength and are limited in pulling, torsional, and jarring strain.
Note: Some sizes of overshots and washpipe may not be available.

Large drill collars may also present a problem if they are too large to form part of a tapered collar string. Too much change in size from large collars to drill pipe or from one size of drill collar to a much smaller size causes accelerated connection failures and rapid fatigue damage to the drill stem. A gradual transition is almost essential to reduce drill pipe fatigue resulting from the change in stiffness.

A good transition can be achieved by: (1) reducing drill collar sizes in the upper end of the drill collar string, or (2) using heavy-walled pipe at the bottom of the drill pipe string, or (3) a combination of the two. Unfortunately, no standard exists that describes exactly how much stiffness change is acceptable. However, computer models can predict the theoretical values to expect from various combinations.

Generally, drill collar size should never be reduced more than 2 in. (50.8 mm) in diameter at any crossover, nor should more than one connection size be reduced at a time. For example, reduction from a 10-in. (254-mm) OD to an 8-in. (203.2-mm) OD to a 6-in. (152.4-mm) OD collar is acceptable; reduction from an 8⅝-in. (219.8-mm) API regular connection to a 7⅝-in. (193.7-mm) API regular, then to a 6⅝-in. (169.6-mm) API regular is also acceptable. At least one 3-collar stand of the next smaller size in each size reduction should be run (fig. 46).

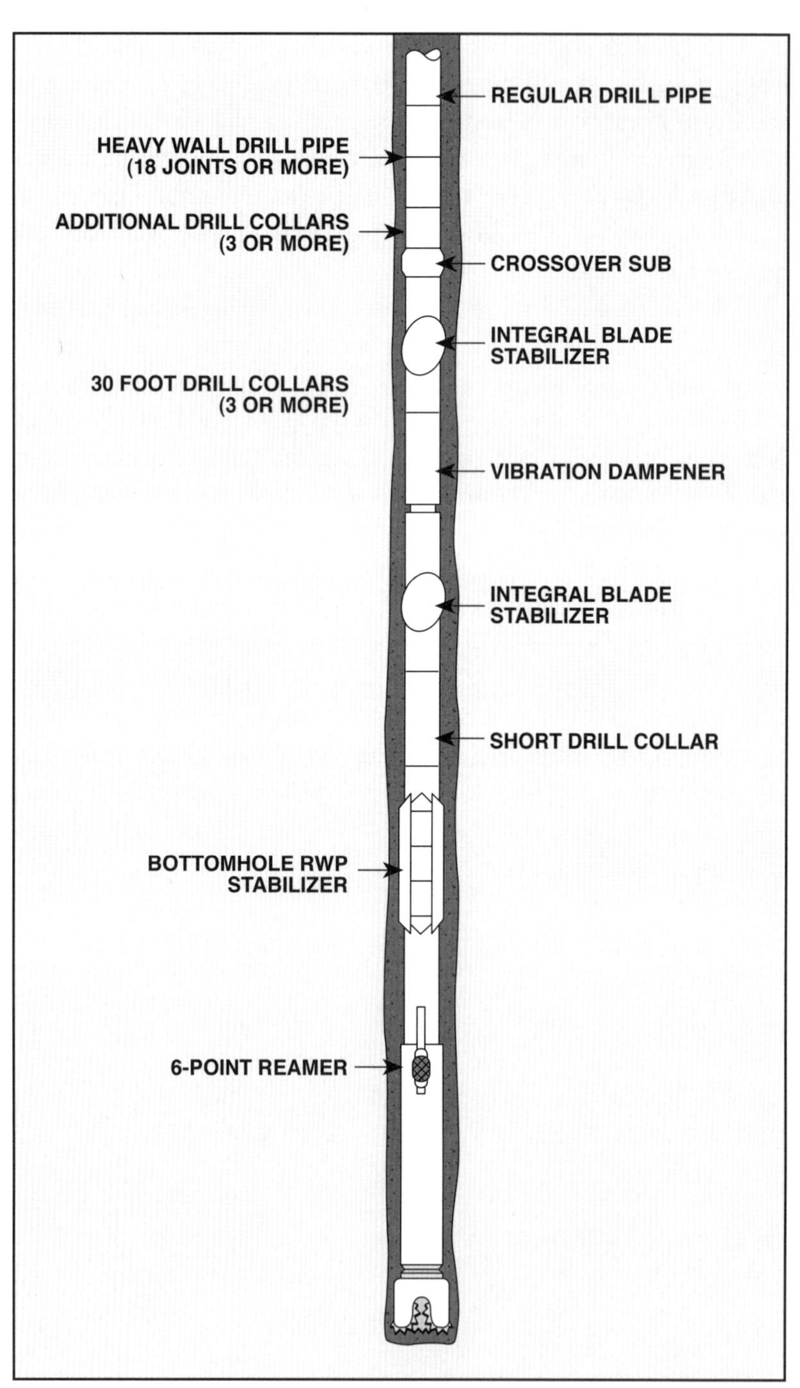

Figure 46. A transitional drilling assembly with reamers, stabilizer, short drill collar, vibration dampener, and heavyweight drill pipe between the drill collars and regular drill pipe (Courtesy of Drilco-Grant Drilling Handbook, Smith International)

Heavy-walled and Heavyweight Drill Pipe

Two types of heavy-walled drill pipe are available for transition sections: (1) conventional heavy-walled drill pipe with a center wear pad; and (2) spiral heavy-walled drill pipe (fig. 47). Both types are equipped with extra-length tool joints that reduce wear by keeping the wall of the pipe away from the side of the hole. Also available is heavyweight drill pipe, which is like conventional drill pipe except that it has thicker walls. For example, manufacturers make a special transition zone drill pipe that has a wall thickness of 1 in. (25.4 mm). By comparison, the wall thickness of regular drill pipe is less than ½ in. (13 mm).

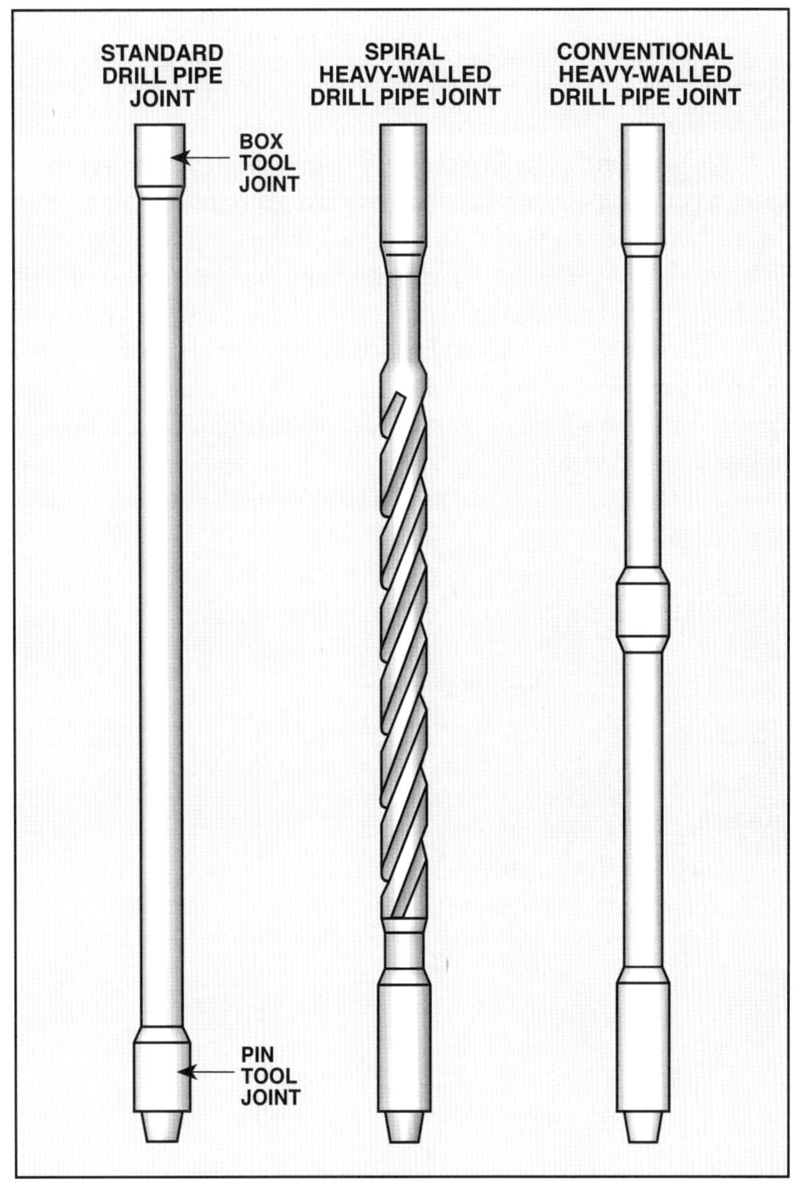

Figure 47. Heavy-walled drill pipe compared to standard drill pipe; note longer tool joint sections.

Calculation of Drill Collar Weight

Drill collar weight must be calculated to include enough reserve so that the drill pipe does not buckle. Drill pipe is subject to serious damage if run in compression. To assure that the drill pipe is always in tension, the usual practice is to use a safety factor of 10 to 25 percent more drill collar weight than the amount needed to keep the drill pipe in tension.

Just as a boat floats in water, drill collars float (are buoyant) in drilling mud. Consequently, drill collars weigh less in drilling mud than in air. The heavier the mud, the greater is the buoyancy effect and the lighter the apparent weight of the collars. Drillers can obtain mud buoyancy factors from tables published by API and pipe manufacturers.

Muds with different densities have different buoyancy factors. For example, a 12-ppg mud has a buoyancy factor of 0.817, which means that a 1-lb weight submerged in the 12-ppg mud would have an apparent weight of 0.817 lb. Likewise, if a 1-lb weight is submerged in 14-ppg mud, it will have an apparent weight of 0.786 lb; the buoyancy factor for the 14-ppg mud is 0.786 (table 6).

In SI units, a 1,487.84 kilograms-per-cubic-metre (kg/m^3) mud has a buoyancy factor of 0.817, which means that the weight of the string in air must be multiplied by 0.817 to obtain its weight when submerged in the mud. Thus, a drill collar that weighs 10,000 kg in air weighs 8,170 kg when suspended in a borehole full of mud. A 1,677.48-kg/m^3-mud has a buoyancy factor of 0.786; thus, a 10,000-kg collar would only weigh 7,860 kg in mud of this weight.

The following equation can be used to determine the necessary weight of drill collars to obtain a desired weight on bit:

$$DC_{wa} = \frac{B_w \times F_s}{F_{mb}}$$

where

DC_{wa} = weight of drill collars in air, lb
B_w = bit weight, lb
F_s = safety factor
F_{mb} = mud buoyancy factor.

Table 6
Buoyancy Factors for Various Mud Weights

Mud (lb/gal)	Weight (lb/ft3)	g/cc or sp gr	Buoyancy Correction Factor
8.34	62.3	1.00	.873
9	67.3	1.08	.862
10	74.8	1.20	.847
11	82.3	1.32	.832
12	89.8	1.44	.817
13	97.2	1.56	.801
14	104.7	1.68	.786
15	112.2	1.80	.771
16	119.7	1.92	.755
17	127.2	2.04	.740
18	134.6	2.16	.725
19	142.1	2.28	.710
20	149.6	2.40	.694
21	157.1	2.52	.679
22	164.6	2.64	.664
23	172.1	2.76	.649
24	179.5	2.88	.633

$$BF = 1 - \frac{\text{Mud lb/gal}}{65.5}$$

Courtesy of Drilco Grant Drilling Handbook, Smith International, Inc.

The safety factor is the percentage of excess drill collar weight used to keep the drill pipe in tension. A 15 percent safety factor is often used; it is written as 1.15. As an example, assume that the required weight on bit is 55,000 pounds and a 12-ppg mud is in use. What will the weight of the drill collar string in air have to be to obtain the required weight on bit? In English units:

$$DC_{wa} = \frac{55{,}000 \times 1.15}{0.817}$$

$$DC_{wa} = 63{,}250 \div 0.817$$

$$DC_{wa} = 77{,}417$$

In this case, the weight of the drill collars in air has to be 77,417 lb to obtain the necessary weight on bit when the drill collars are run into the borehole full of mud.

$$DC_{wa} = \frac{24{,}475 \times 1.15}{0.817}$$

$$DC_{wa} = 28{,}146 \div 0.817$$

$$DC_{wa} = 34{,}451$$

In this case, the weight of the drill collars in air has to be 34,451 dN to obtain the necessary weight on bit when the drill collars are run into the borehole full of mud.

Drillers can calculate the number of drill collars required to obtain the necessary weight on bit by using tables provided by manufacturers of drill collars. The tables give weights of different-sized drill collars in lb/ft (see table 4). (They are also available in kg/m. Also, you can convert lb/ft to kg/m by multiplying lb/ft by 1.488.) Using the information from these tables, you can determine that the required weight on bit in the previous example could be obtained by using an assembly of nine 8-in. OD by 2 13⁄16-in. ID collars and twelve 6¾-in. OD by 2 13⁄16-in. ID collars. In SI metric terms, an assembly of nine 203.2-mm by 71.4-mm ID collars and twelve 171.5-mm OD by 71.4-mm accomplishes the required weight.

Square Drill Collars

A square drill collar is a square bar with rounded corners (fig. 48). The wall contact areas of the square collar are usually coated with tungsten carbide to resist wear. Fluid is circulated between the flat sides of the collar and the wall of the hole. Round surfaces near the ends of the collar provide space for tonging.

A square drill collar is an excellent tool for retarding the change of angle deviation in a well. The corners of the square collar come into contact with the wall of the hole and, if the formation is hard enough, the hole wall supports the collar to keep it from flexing. Although square drill collars cannot completely prevent angle from building up, they can prevent it from building at a rapid rate, thereby preventing doglegs. Square drill collars are less effective when used in soft formations because the soft formations do not support the collars well enough to allow them to retard angle buildup. When total angle must be reduced, operators normally remove square drill collars from the string and run a pendulum assembly consisting of round collars and stabilizers.

Figure 48. Square drill collar

Because a new square collar is only 1⁄16 in. (1.5 mm) smaller than the hole size, it will stick in the hole or its corners will wear excessively if the bit wears undergauge. If the collar wears as much as 3⁄16 in. (5 mm) smaller than hole diameter, it will lose its effectiveness. To protect it, operators usually run a rolling reamer-stabilizer above the bit (see fig. 37). The reamer helps reduce wear on the collar and prevents dragging and torquing on the string when the bit begins to wear undergauge. The diameter of a square collar is measured between opposite corners (fig. 49).

A nonrotating stabilizer just above the square collar centers the collar in the hole and reduces wear. Because clearance between the square collar and the wall of the hole is such a critical factor in the effectiveness of the collar, the collar should be gauged every trip and replaced if necessary. Square collars can be rebuilt by controlled metallurgical procedures.

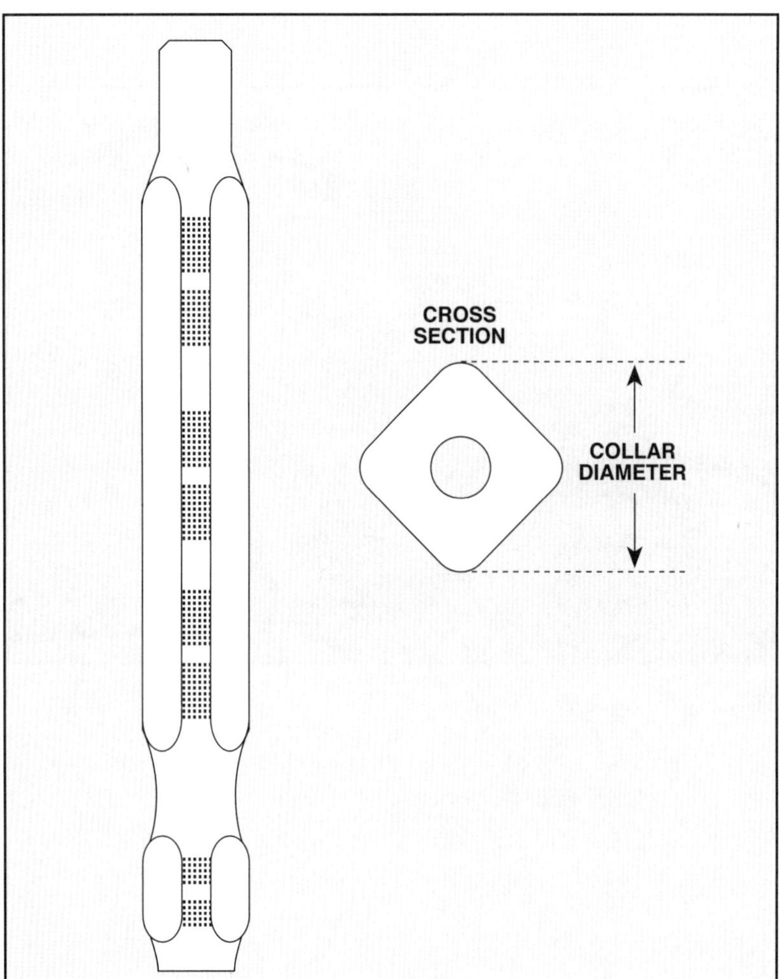

Figure 49. Square drill collar drawing showing cross section diameter

Spiral Drill Collars

Spiral drill collars (fig. 50) are used in holes where the clearance between the hole and the drill collars is small and where the collars will likely be in contact with the borehole wall, a situation that can result in differential sticking (fig. 51). Differential sticking is a condition in which the drill stem becomes stuck against the side of the borehole because the drilling fluid pressure is greater than the formation pressure. This difference in pressure forces the liquid part of the drilling fluid (the filtrate) into a permeable formation and causes a buildup of mud solids (filter cake) on the wall. Filter cake reduces hole size, which allows the drill collar to come in contact with the filter cake. Continued flow of filtrate into the permeable formation creates a suction that can hold the collar against the side of the hole. Spiral or grooved drill collars reduce the drill collar surface that can actually contact the wall. This reduced wall contact lessens the possibility of differential sticking.

Figure 50. Spiral drill collars

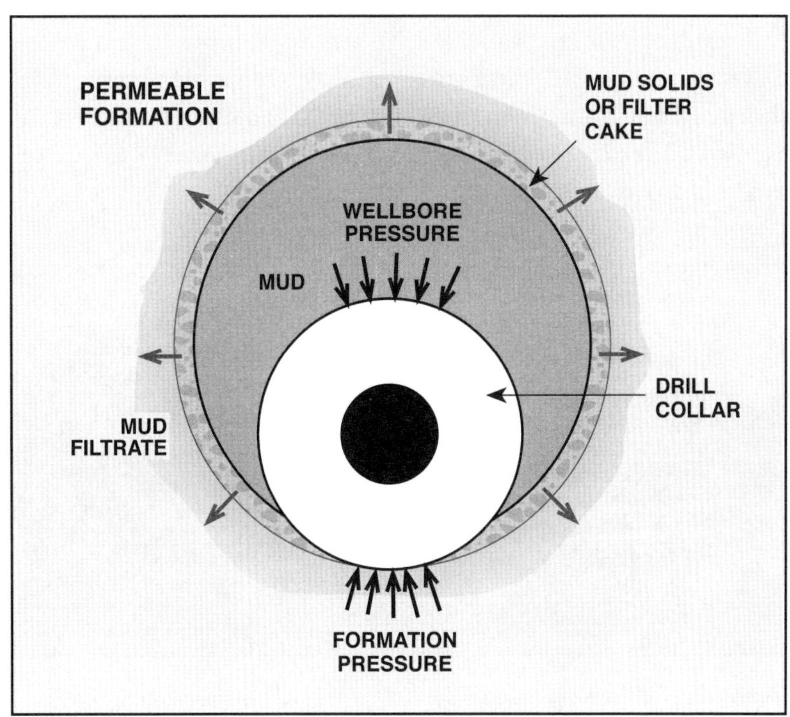

Figure 51. Differential sticking

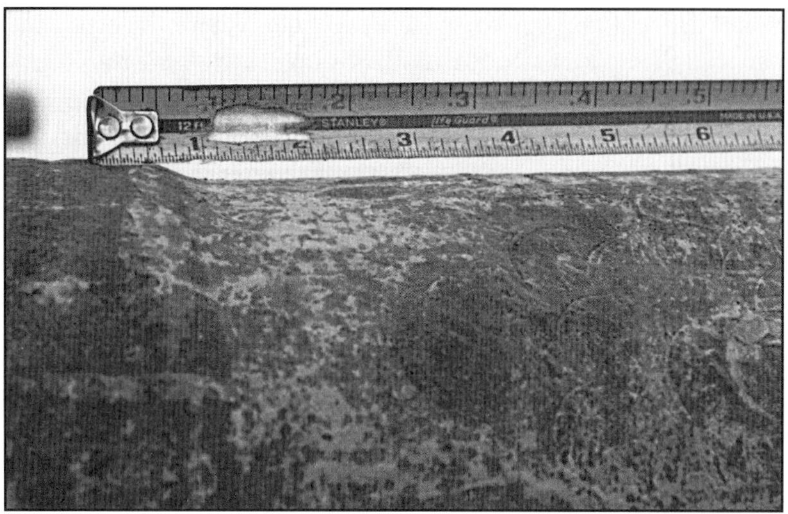

Figure 52. Drill collar with recess for use with slips and elevators

Figure 53. Hardbanding on a drill collar for use in hard abrasive formations

Drill collars are available with special features, including recesses for slips and elevators (fig. 52) and hardbanding where hard abrasive formations are present (fig. 53). Nonmagnetic collars are used when magnetic deviation and directional surveys are to be run. They may be either flexed or slick (fig. 54).

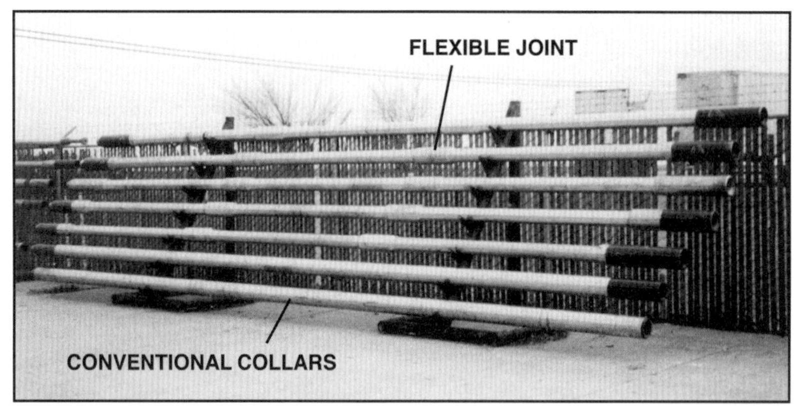

Figure 54. Nonmagnetic drill collars are used for running magnetic deviation or direction surveys. The five on top have flexible joints, which allow the collar to bend when drilling directional holes. The two on the bottom are conventional collars used in straight hole drilling.

Stabilizers

Stabilizers are an integral part of a BHA; they stabilize the bit and the drill collars in the hole. When a BHA is properly stabilized, optimum drilling weight can be applied to the bit (fig. 55). Further, proper stabilization combined with optimum weight forces the bit to rotate on its true axis. A bit rotating on its true axis loads the bit cones equally, which allows the bit to drill straight ahead without sudden changes. Fewer bits will be used and the penetration rate will increase.

For a BHA to maintain a straight course, stabilizer blades must be as near bit size as possible. The most important consideration when choosing the right stabilizer for a well is the type of formations that will be penetrated. The stabilizer must be durable; the harder and more abrasive a formation is, the more durable the stabilizer must be.

The stabilizer must also have adequate surface area in contact with the wall of the hole to prevent it from digging into the wall and losing its effectiveness. In a hard formation, a stabilizer with a small wall contact area performs satisfactorily, but a larger wall contact area is required in a softer formation.

Figure 55. Rotating stabilizers with spiral blades

Total wall contact area of stabilizers near the bit must increase as crooked-hole tendencies become more severe. Using stabilizers with longer and wider blades and adding more stabilizers increases the wall contact area. However, increasing the blade area decreases the fluid passage area around the stabilizers. Too much fluid restriction can cause sticking if cuttings pack off around the stabilizers and lost circulation if too much back-pressure builds up on the formation. Too much restriction can also cause erosion in certain types of formations. Therefore, operators must take care to select a stabilizer that provides adequate wall contact area without excessively restricting annular flow.

Stabilizers can be divided into three categories: (1) rotating blade, (2) nonrotating sleeve, and (3) rolling-cutter reamer (fig. 56).

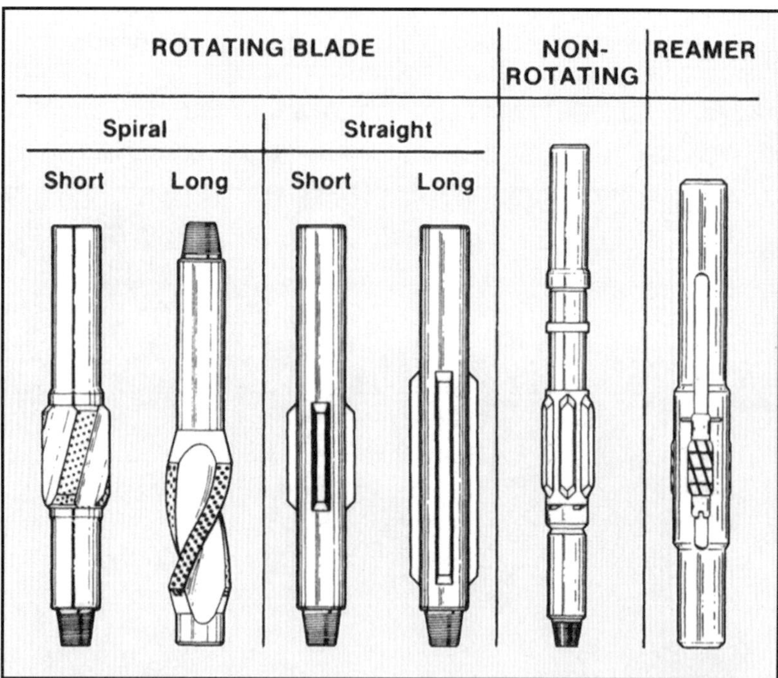

Figure 56. Three basic categories of stabilizing tools (Courtesy of Smith International, Inc.)

Rotating-blade Stabilizer

A rotating-blade stabilizer can have straight or spiral blades and the blades may be short or long. All rotating-blade stabilizers have fairly good reaming ability, and because of recent improvements in hardfacing, they have very good wear life. Hardfacing materials include granular tungsten carbide, crushed sintered tungsten carbide, inlaid sintered tungsten carbide, and pressed-in sintered tungsten carbide inserts.

Rotating-blade stabilizers can be divided into two groups: those that can be repaired at the rig site and those that must be repaired at a service location. Welded-blade, integral-blade, and shrunk-on sleeve stabilizers must be repaired or reconstructed at a service location, while most sleeve stabilizers and replaceable-blade or wear-pad stabilizers can be repaired on the rig.

Welded-blade Stabilizers

Welded-blade stabilizers are most frequently used in soft and medium formations (fig. 57A). Operators could not use the earliest designs in hard formations because of the rapid fatigue damage that occurred in the areas where blades were welded in place. Also, they could not stay in gauge long enough to make them practical in hard formations. New welding and hardfacing techniques have improved welded-blade stabilizer performance and operators now use them in most areas, even in hard rocks. A variety of blade configurations is available for all borehole sizes.

Integral-blade Stabilizers

Integral-blade stabilizers are extremely durable and rugged; consequently, operators use them in hard and very hard formations (fig. 57B). They are constructed from a single piece of high-strength alloy steel, and blade faces are hardfaced with pressed-in tungsten carbide inserts. Cloverleaf or spiral blade configurations are available.

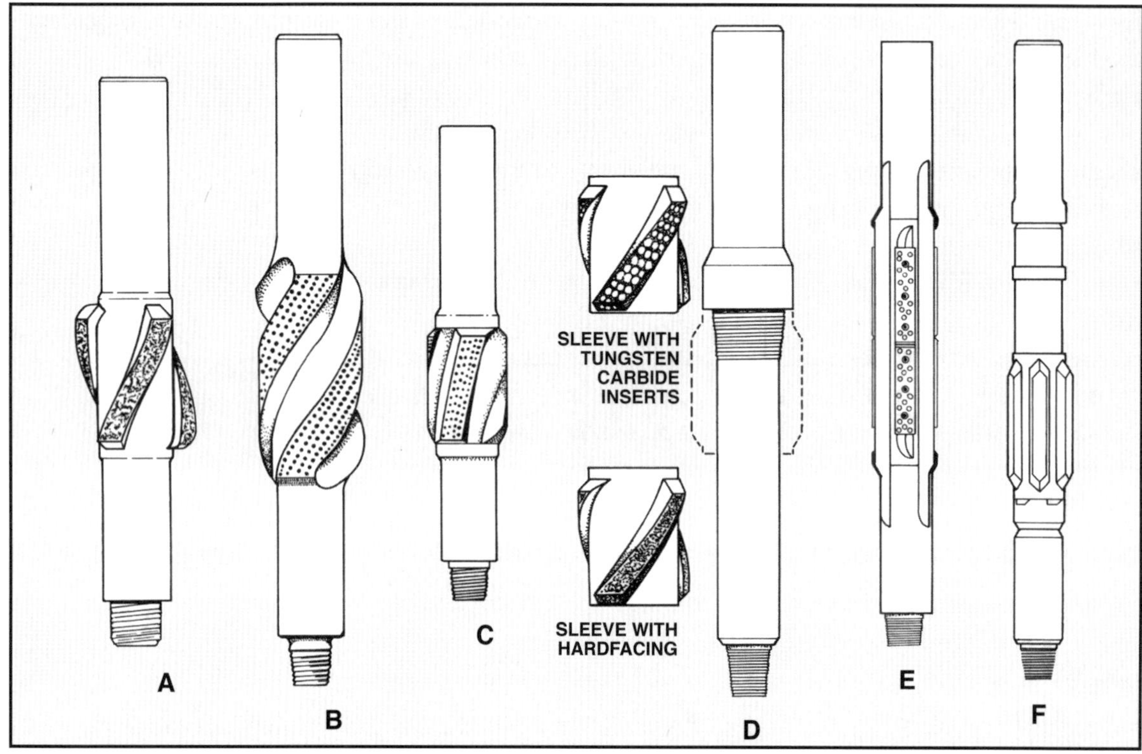

Figure 57. Types of stabilizers: A. welded-blade; B. integral-blade; C. shrunk-on sleeve; D. two types of threaded sleeves; E. replaceable rollers; F. nonrotating rubber sleeve

Shrunk-on Sleeve Stabilizer

The shrunk-on sleeve stabilizer is a type of integral stabilizer constructed of two pieces: a body and a sleeve (fig. 57C). During assembly of the stabilizer, the sleeve is heated (up to 750°F or 400°C) to expand the bore and is then cooled so that it contracts around the body.

Blades are hardfaced with crushed tungsten carbide or pressed-in inserts. When the blades are worn out, the sleeve is removed with a cutting torch and a new sleeve is installed. The same body can be used for several hole sizes.

Threaded-sleeve Stabilizer

A threaded-sleeve stabilizer can be used in all but the hardest formations (fig. 57D). It is comprised of a one-piece body and a replaceable screw-on sleeve with blades. A single body can be used with various sleeves for different hole sizes. The entire sleeve must be replaced in order to change blades, but it is easily made up on the rig floor with tongs or a sleeve breaker.

Replaceable-blade Stabilizer

The replaceable-blade (or wear-pad) stabilizer is most frequently used near the bit, when maintaining hole gauge is important, and in hard or abrasive formations (fig. 57E). The body of the stabilizer is machined to hold replaceable blades that are available in wedge or pad designs. The blades are held in place by bolts and may easily be changed on the rig floor with hand tools. Some blades contain wear indicators that specify the exact amount of downhole wear.

Nonrotating Sleeve Stabilizer

The nonrotating-sleeve stabilizer is most effective in hard formations, such as limestone and dolomite (fig. 57F). It has a ribbed rubber sleeve mounted on a mandrel body. During drilling operations, the sleeve remains stationary in the hole, while the mandrel and drill collars rotate. The stabilizer thus acts as a drill bushing. Because the sleeve does not rotate, it keeps hole gauge far longer than hard-surfaced stabilizers and does not damage the wall of the hole. The sleeve reduces drill collar whip, which increases stability and reduces connection fatigue. Sleeves can be replaced at the rig site. The rubber sleeve does have some limitations, however. It has no reaming ability and cannot be used where temperatures exceed 250°F (120°C). It also has short wear life in holes with rough walls.

Rolling-cutter Reamer

A reamer is run between the bit and the drill collars. The reamer body rotates while the rolling cutters open the hole to full gauge, thereby extending bit life and preventing problems with sticking. It is primarily used for maintaining full-gauge hole in hard formations and is the only tool that can effectively do so, even though its wall contact area is very small.

A reamer is also used for additional stabilization in hard formations. In most instances, however, it is a poor stabilizer because of its very limited wall contact area. If used in soft formations, its cutters will penetrate the wall of the hole, causing reduced stabilization and bit deviation.

The bit will also deviate if an unstabilized reamer is used while drilling with high weights. Operators should run at least one stabilizer directly on top of the reamer in medium crooked-hole conditions, and should run a minimum of two stabilizers above the reamer in severe crooked-hole conditions.

A reamer body is constructed of high-strength alloy steel. A basic reamer is equipped with three cutters and is called a three-point reamer (fig. 58A). A six-point reamer is also available and is used in extreme conditions, where more wall contact is needed for stabilization or more cutting action is required. The six-point reamer consists of two three-point reamers in a single body (fig. 58B). Cutters are mounted vertically or at a slant. Vertical cutters give a true roller action, while slanted cutters give a scraping, gouging action.

Various cutters are available for use in different formations (fig. 59). They can be changed easily on the rig floor when they have become worn or undergauge. When a new cutter is installed, the smaller end is placed downward so that if the bit drills undergauge, the end of the cutter will fit the hole made by the bit.

Like stabilizers, reamers must be maintained at a diameter as near bit size as possible to work effectively. Bear in mind that installing a reamer directly above the bit nullifies the benefits of the pendulum technique (fig. 60).

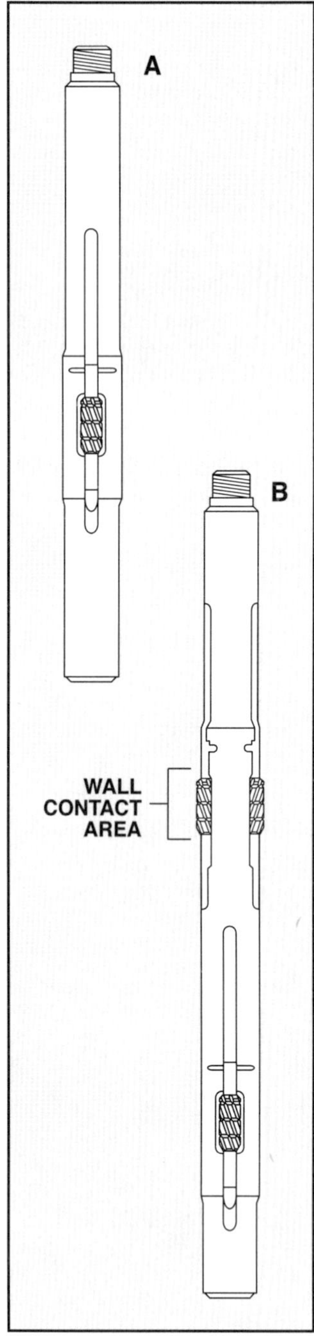

Figure 58. Roller reamers: A. three-point reamer; B. six-point reamer

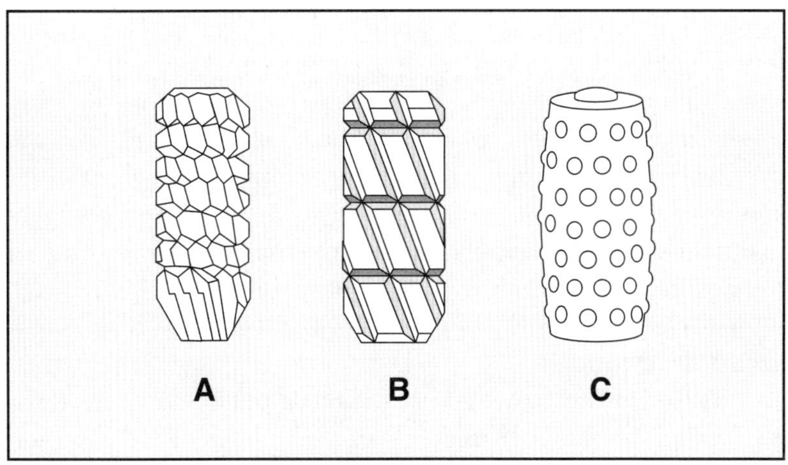

Figure 59. Three types of cutters for roller reamers: A. hardfaced sharp teeth for soft formations; B. hardfaced flat teeth for medium-hard formations; C. tungsten carbide inserts for hard formations

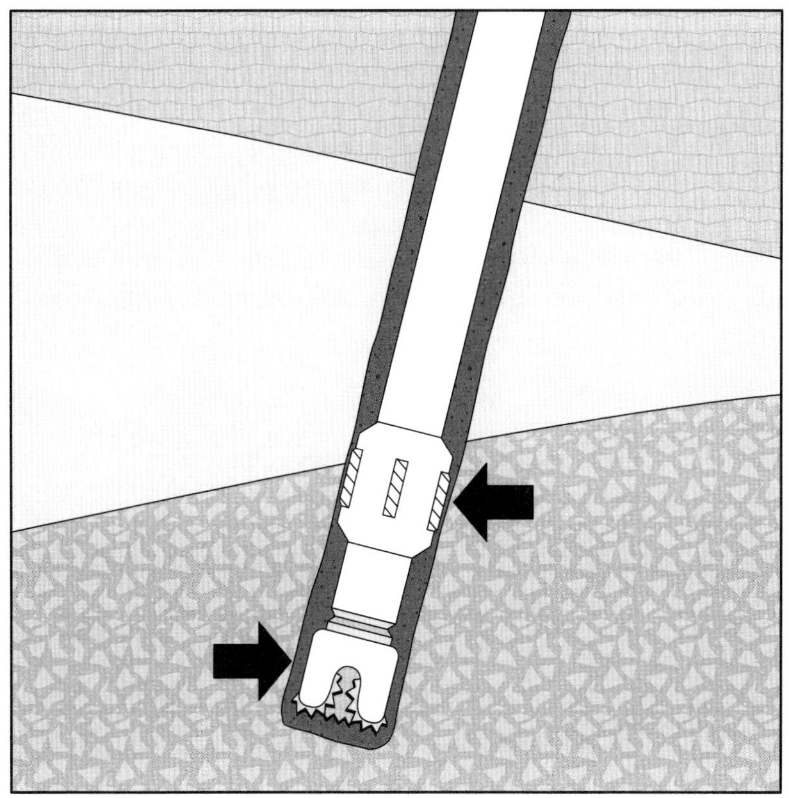

Figure 60. A reamer run above the bit to insure a full-gauge hole will nullify the pendulum effect. (Courtesy of Smith International, Inc.)

Vibration Dampeners

A vibration dampener (a shock sub, or shock absorber) is designed to work like a shock absorber; it permits normal drilling without subjecting the drill stem to the damaging bounce and vibration produced by the roller cone bit as the cones rotate. A vibration dampener may contain rubber, springs, compressed gas, or other springing elements for absorbing shock (fig. 61). The shock absorber maintains a constant bit pressure on the rock. By eliminating bit bounce and the tendency to walk, penetration rates are enhanced, bit life is lengthened, the hole is straighter, and there is less stress on the drill string.

When drilling the upper portion of the hole, the need for a vibration dampener is greater because bit bounce and vibration are readily transmitted to the surface. When the well gets deeper, however, bounce and vibration may not be noticeable at the surface because of the dampening effect of the long drill stem. Small changes in rotary speed, bit weight, or formation can cause excessive gyrations at the bottom of the hole. Modern MWD tools can record such downhole drilling conditions and alert the driller to the need for a vibration dampener. Close inspection of dull bits and routine inspection of the drill stem can detect damage from rough drilling. Also, some formations are known for rough drilling characteristics that require a shock sub in the drilling assembly.

To effectively absorb shock and protect the drill stem, the ideal position for a vibration dampener is directly above the bit. When a PHA is used, the ideal location is above zone 3, with an additional stabilizer run 30 ft (9 m) above it to prevent excessive lateral loading.

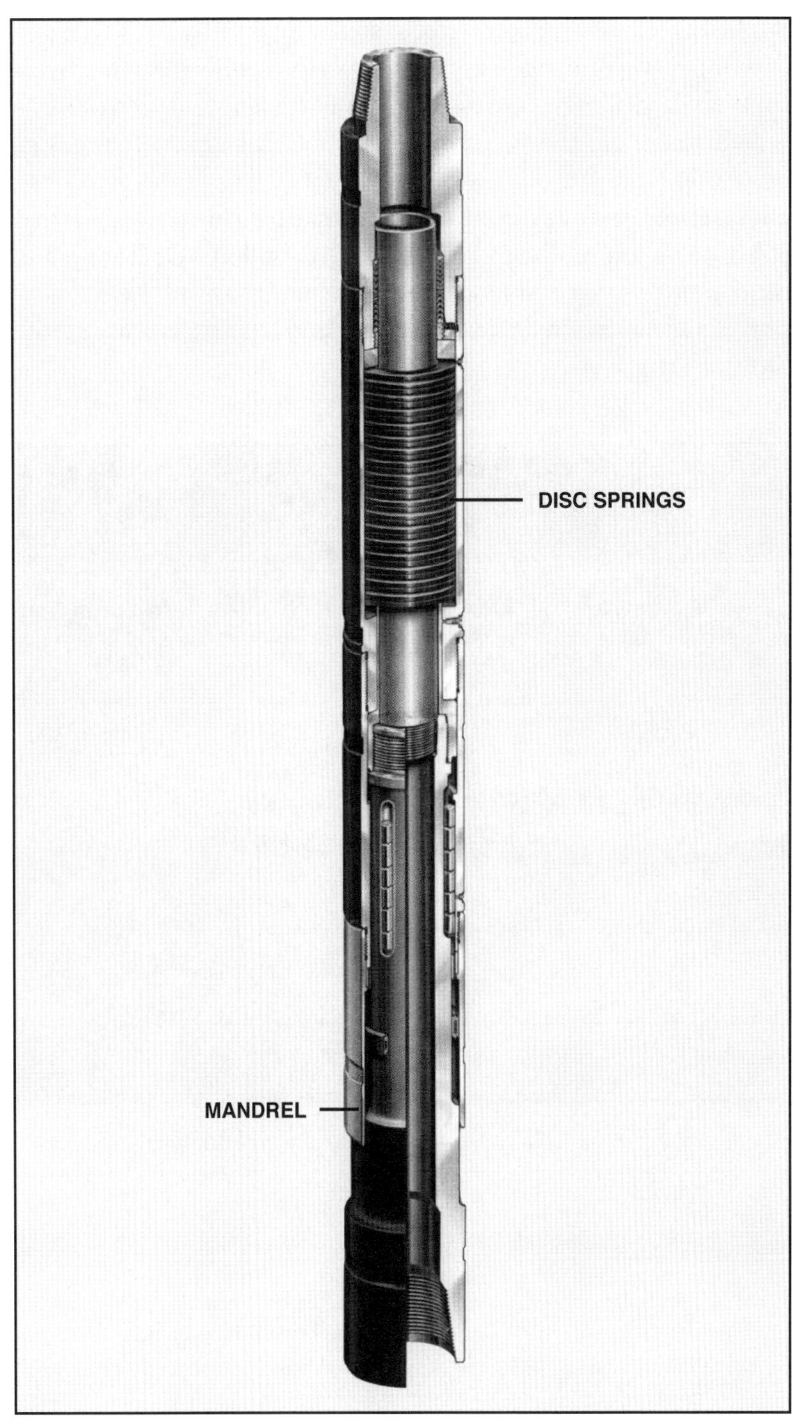

Figure 61. Cutaway drawing of a vibration dampener tool. Also referred to as shock sub or simply a shock absorber. Size ranges from 4¾ to 10 inches (120.7 to 254 millimetres) in diameter; 9 to 12 feet (2.7 to 3.7 metres) in length. (Courtesy of Security DBS, A Halliburton Company)

Measurement while Drilling (MWD) Tools

Sophisticated rig site computer systems, tied to both surface and downhole instruments, can provide critical information in real time while drilling progresses. Data is transmitted from downhole detectors to the surface by mud pulse telemetry (fig. 62) or by wireline (in downhole drilling motors). Rock type, hardness, compressive strength, and other formation characteristics that affect deviation can be monitored at the rig floor. Drill stem and bit conditions that influence rate of penetration and hole deviation are instantly available for review both on site and at remote locations (fig. 63).

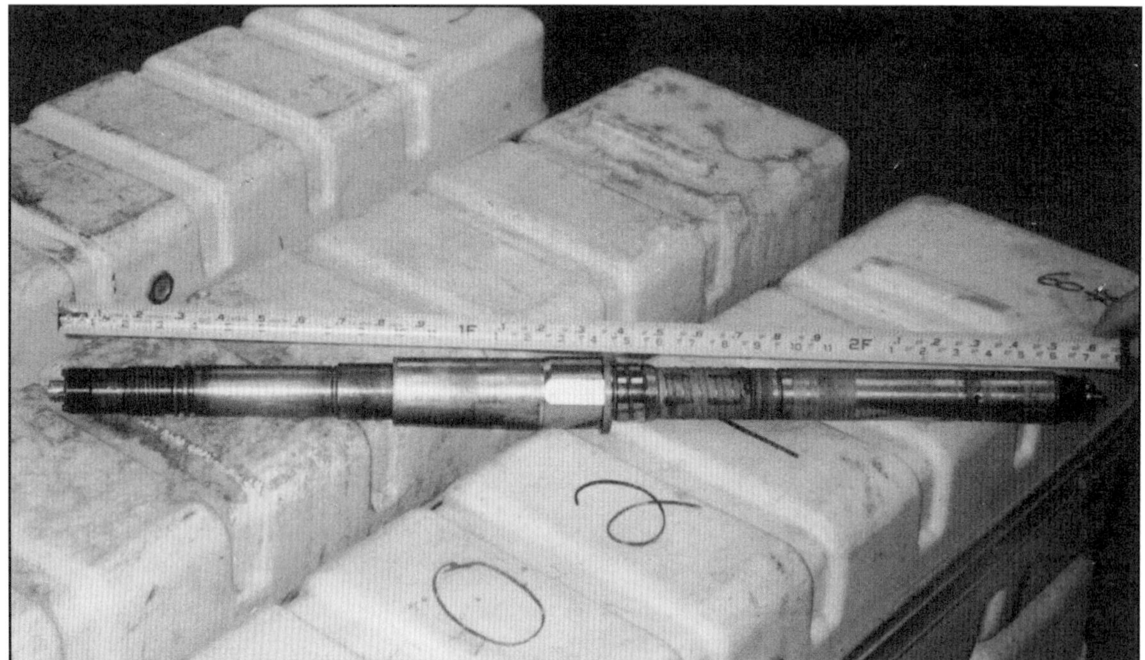

Figure 62. Downhole positive pulse instrument for MWD transmits data through the mud column.

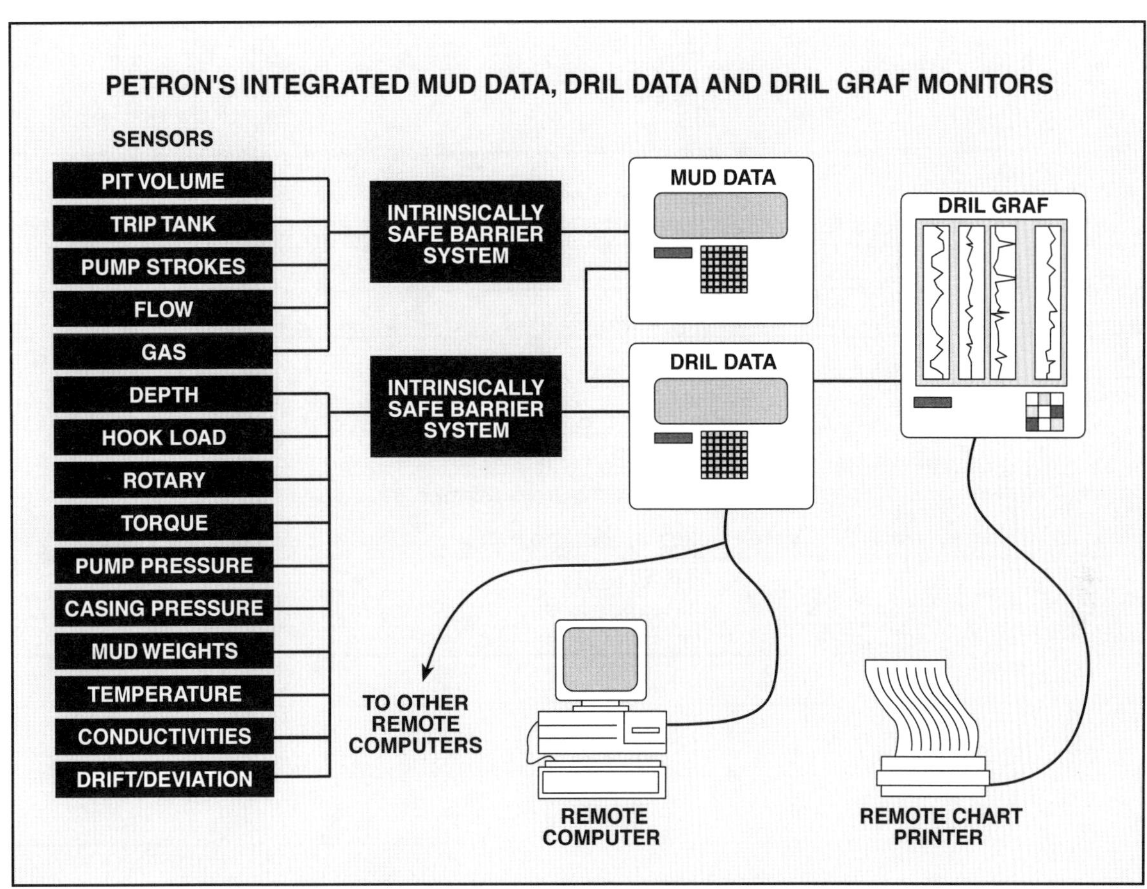

Figure 63. A rig site integrated drilling system for measurement while drilling

---

To summarize—

Drill stem tools used in straight-hole drilling include

- standard drill collars
- heavy-walled and heavyweight drill pipe
- square drill collars
- spiral drill collars
- stabilizers
  1. rotating blade
  2. nonrotating sleeve
  3. rolling-cutter reamer
- vibration dampeners
- MWD tools

Deviation-Recording Instruments

The contract between the contractor and the operating company specifies the maximum deviation a drilling contractor is allowed. The contract may also state how often a deviation survey is to be run. This frequency can vary, in terms of footage (or metreage) drilled, from 15 to 500 ft (5 to 50 m), depending on the area to be drilled (see fig. 8). Intervals of 100 to 250 ft (30 to 75 m) are normal in areas where drilling is slow and few deviation problems occur. If deviation problems are expected, contractors may be required to run a survey more often or, for their own protection, they may decide to run a survey more often. Frequent surveys are desirable but they are also expensive. Although a majority of contractors own deviation-recording instruments, the cost in downtime required to lower and retrieve the instruments can become excessive if too many surveys are run.

When measurements are taken, any deviation must be recorded and compared with the amount permissible in that part of the hole. Supervising personnel must be completely informed of all survey results.

Deviation-recording instruments measure only deviation, or drift (inclination from the vertical). They do not measure the direction of the deviation; they do not record compass headings. More sophisticated instruments are required to measure hole angle and direction. Such instruments are used in directional drilling to record the drift and the direction of the wellbore.

MWD computer systems can provide instant drift and direction information where the rig is so equipped. However, most deviation information, especially on standard rotary drilling operations, is still obtained by running deviation-recording instruments down the drill string.

Running Methods

A primary consideration for the driller when running and retrieving a deviation-recording instrument is the length of time required for the operation. When a survey is being recorded, the drill stem must be kept still for a certain length of time. The longer the drill stem is at rest, the greater is the chance that it will become stuck. Drillers need an accurate timing device on the surface that is synchronized with the timing device in the instrument so that he will keep the pipe motionless for a minimum amount of time. Newer tools use an automated timer system, eliminating the clock-setting-and-make-up time.

If an instrument has been properly maintained, the time required to set the clock and make up the go-devil—the device in which the instrument is run—should be 1 to 1½ minutes (min). Timing is important because while the tool is being made up, the drill pipe is stationary, a time that should be kept to a minimum to avoid stuck pipe.

The time a go-devil takes to reach the point where a survey is to be taken depends on the running method used and the condition of the mud. When the go-devil is dropped freely in a mud-filled drill pipe, calculations are made on an assumed descent rate of 1,000 ft per minute (ft/min) or 100 m/min. While the tool is descending, the driller should slowly rotate the drill string and raise and lower it to avoid sticking. When a tool is run on a sandline or measuring line, operators usually assume a descent rate of 500 ft/min (50 m/min). Again, the driller should raise and lower the drill stem while the tool is descending to prevent sticking. Once the tool is on bottom all movement should stop 30 to 45 seconds (sec) to allow vibration to cease and to permit the indicator to come to rest.

The go-devil assembly protects the deviation-recording element from shock and other damage. Three principal methods are used for running a go-devil (fig. 64):

1. It may be run in and pulled out of the drill pipe on light measuring line or on ordinary sandline.
2. It may be dropped into the drill pipe and retrieved with a core barrel overshot assembly that is attached to an ordinary sandline or measuring line.
3. It may be dropped into the drill pipe and recovered when the string is removed from the hole.

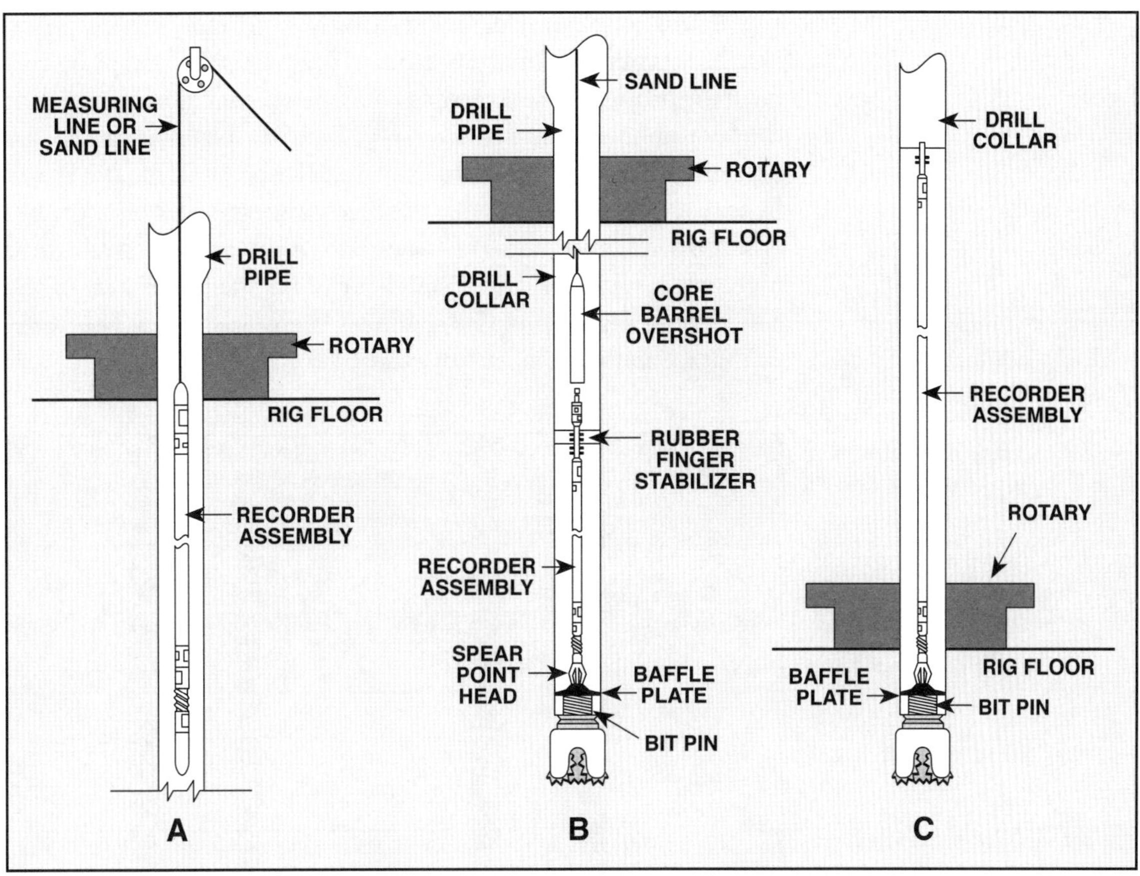

Figure 64. Three methods of running a go-devil deviation survey: A. it is run in and pulled out on a measuring line or a sand line; B. it is dropped into the drill pipe and retrieved with a core-barrel overshot tool; C. it is dropped into the drill pipe and recovered when the string is removed from the hole.

Methods of landing and centering the go-devil assembly in the hole vary according to the methods used for running the assembly. If the go-devil is run in with a measuring line or sandline, it is first lowered to the bit and then raised a few feet so that it lies in the lowest drill collar. In this position, its axis is parallel to the axis of the drill pipe, and an accurate reading can be taken.

If the go-devil is dropped into the drill pipe, a spear point assembly is made up on the bottom of the go-devil. The spear must be of the proper diameter to fit a baffle plate that has been placed in the bottomhole assembly (fig. 65). The baffle plate may be seated on the bit pin, in a tool joint, or in a drill pipe float. Ideally, it will be seated on the bit pin so that the recording can be taken at the bit. The baffle plate centers the lower end of the go-devil in the drill pipe, while a rubber-finger stabilizer holds the upper part of the assembly in alignment. The stabilizer keeps the go-devil centered so that a core barrel overshot can be properly engaged.

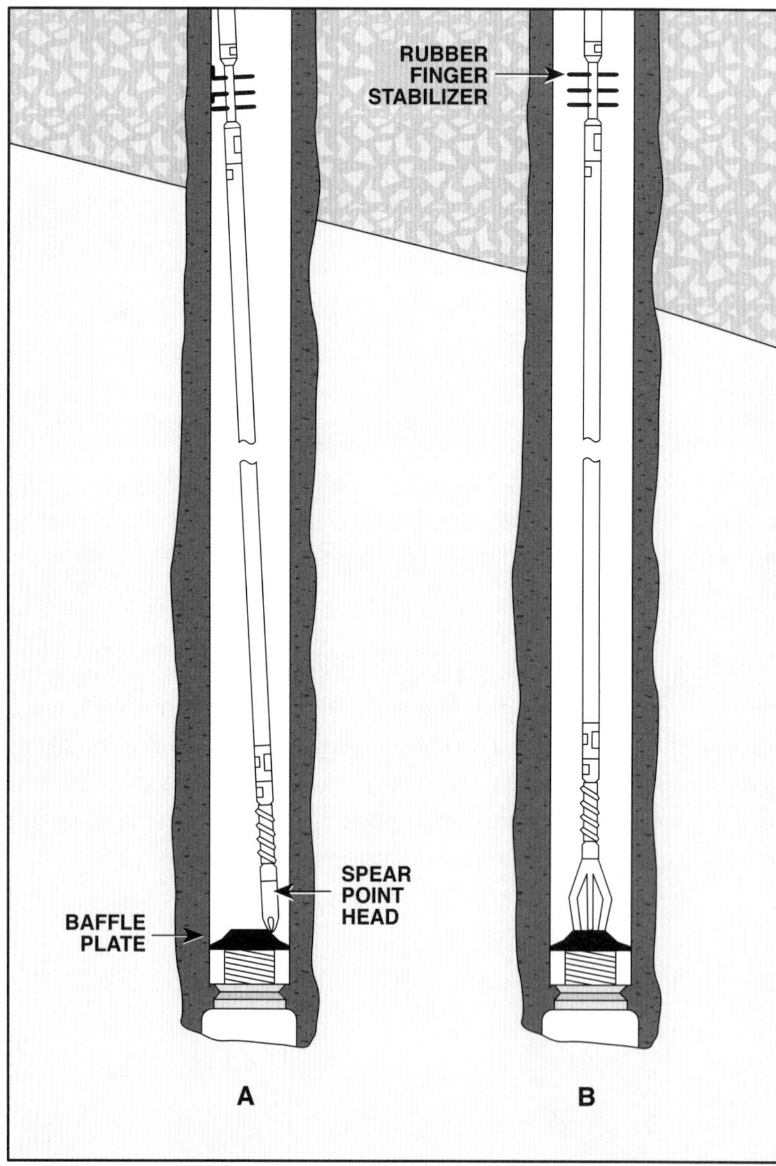

Figure 65. A spear point assembly. If the spear point head is too small, an incorrect record may result (A). *A spear point head of the proper size centers and guides the recorder for a correct reading* (B).

Types of Instruments

Several types of deviation-recording instruments are available for straight-hole drilling. Some measure deviation only; others measure both deviation and direction (azimuth). The basic operating principle used in the instruments is the pendulum effect. The pendulum is on a pivot point and indicates the hole angle on a recording disc. Discs are made of paper, metal, or photographic film. Concentric circles printed on the discs show the angle of inclination, or deviation, from vertical. Directional discs include references to magnetic north. Timing devices are included in the instrument to ensure that the disc will not be perforated or marked until enough time has elapsed for the unit to be bottomed in the drill collar above the bit and come to rest. In any type survey, both the instrument and the drill stem must be perfectly still for an accurate recording. An inaccurate reading will also result if the housing is bent or is improperly centralized (see 'A', fig. 65).

The double recorder is the most commonly used straight-hole deviation recorder (fig. 66). It consists of a recording instrument and a go-devil. Double refers to the process in which the recording disc is punctured, picked up, turned 180 degrees, and punctured again. This process verifies that the go-devil was at rest and landed properly. The time between shots is approximately 25 seconds. Both holes should indicate the same angle; if they do not, the survey is inaccurate.

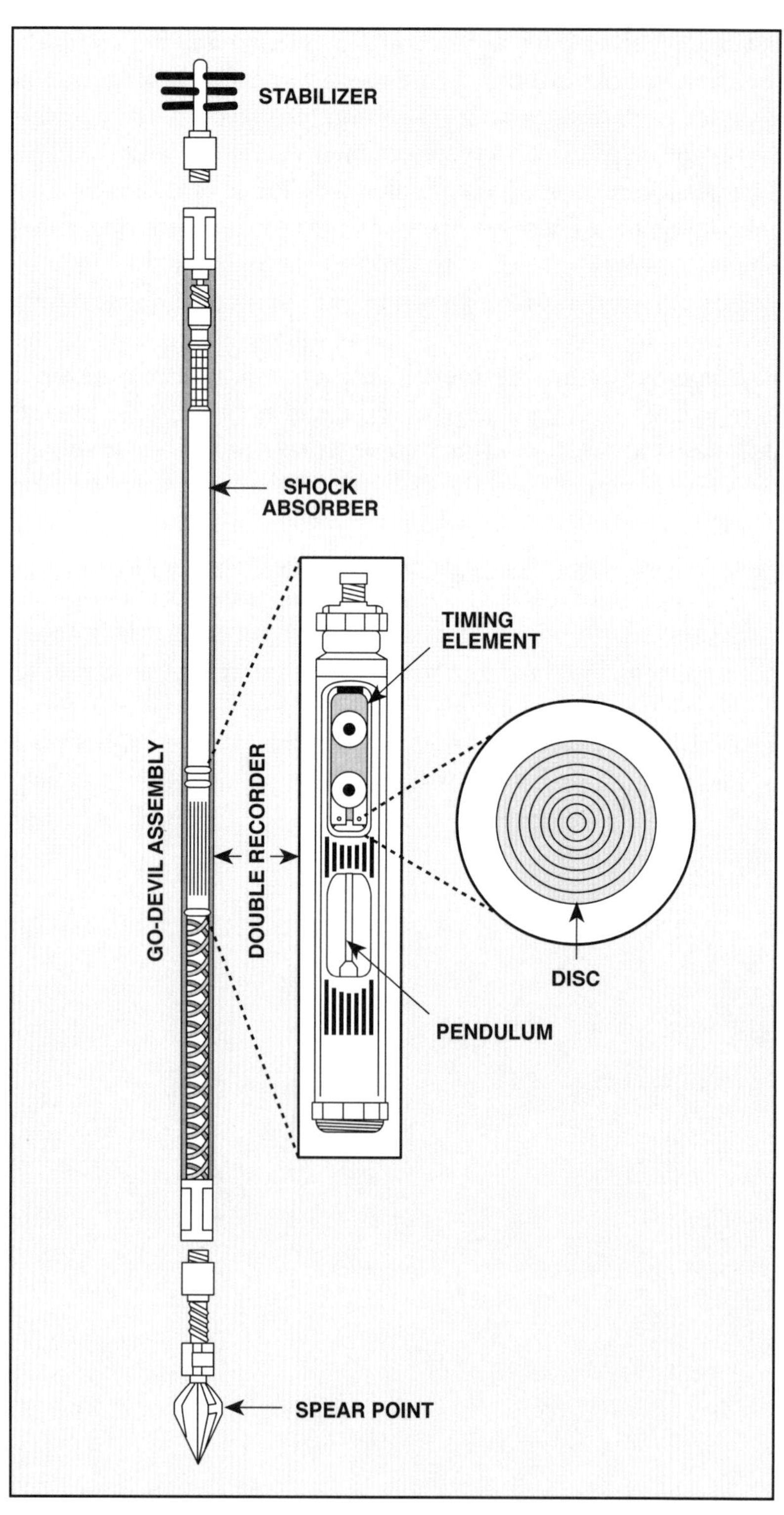

Figure 66. Double recorder disc and go-devil assembly equipped with stabilizer and spear point

Another straight-hole recorder that works on the pendulum principle is the multiple recorder. It measures both deviation and direction by means of a compass or gyroscope built into the tool (fig. 67). It is equipped with a piston built onto the spear point. When the go-devil lands on the baffle plate, the main piston carries the pendulum upward to contact the record disc. When the go-devil is lifted off, the instrument recocks itself, and another recording can be made. Any number of readings can be taken on a single disc as the instrument is picked up and set down on the baffle plate. To make more than one recording, the go-devil must be run on a wireline. If the directional tool uses a magnetic compass, it must be run in nonmagnetic drill collars because steel drill collars affect the accuracy of the compass readings. Gyroscope models can be run in regular drill collars because gyroscopes do not depend on magnetism to obtain direction.

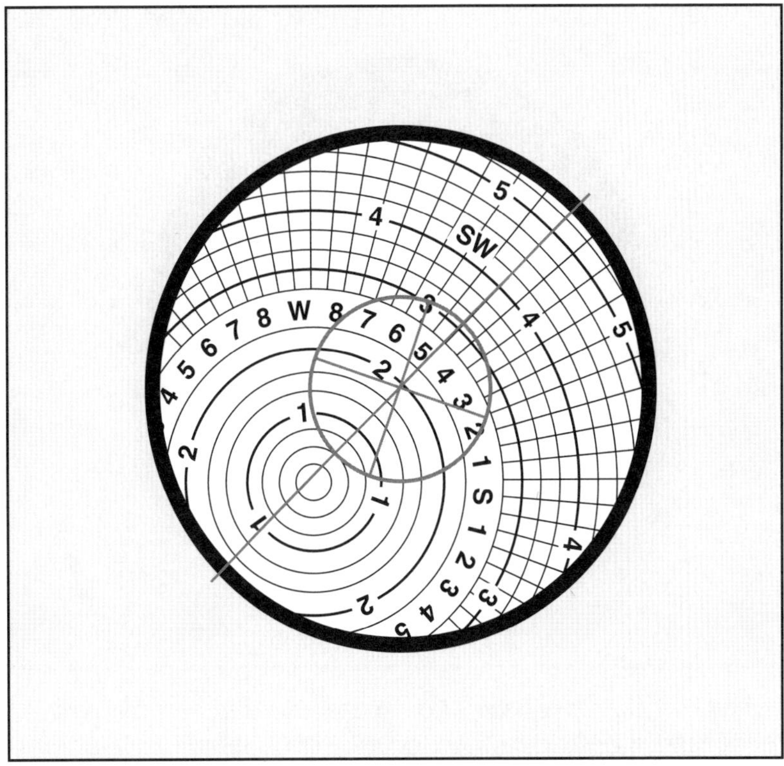

Figure 67. Example of a compass disc showing borehole inclination and direction using a 6° compass. Crosshair indicates 1 ¾° inclination. Line from center of compass through crosshair shows direction of inclination to be S 35° W. (Courtesy of Sperry-Sun Drilling Services, a Halliburton Company)

An inclinometer permits multiple recordings of deviation (but not direction). It uses a small electric current to produce a circular white dot on a chemically treated disc when the inclinometer is at rest. The size of the dot varies with the duration of the still period of the pendulum (fig. 68). The point of the pendulum has a floating stylus that contacts the disc at all times except when the instrument is placed in a horizontal position. A hairspring then draws the stylus into the pendulum so that the electrical current is broken. A wireline operation allows multiple records at different depths with one trip. The dots are identified as to depth by varying the still time at each depth, which in turn varies the size of the dots. The largest dot indicates the most reliable reading because it results from the longest still time, which in turn produces the most stable recording condition.

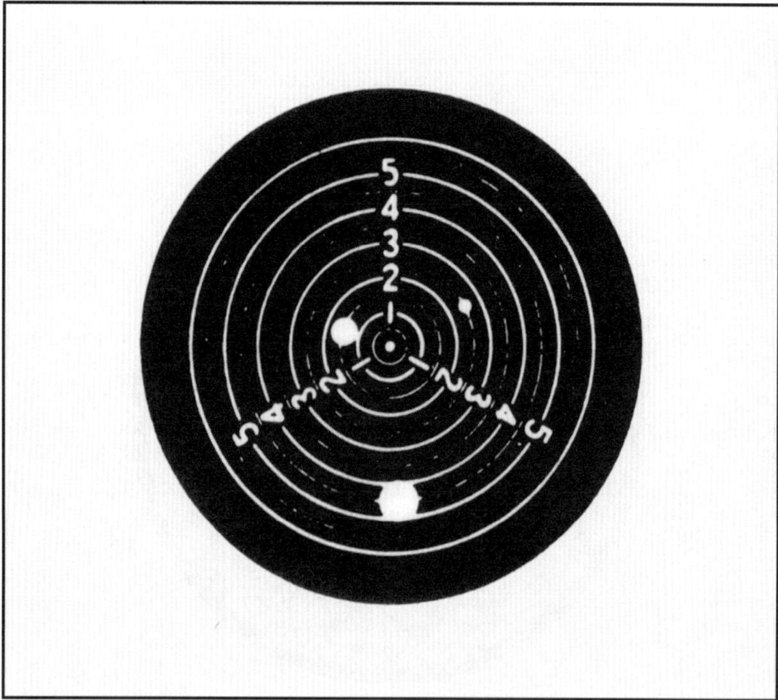

Figure 68. Disc from a multishot inclinometer with three deviation recordings. Large dot is from longest still-time and is the most accurate reading.

To summarize—

Deviation-recording instruments

- measure only deviation, or drift, or measure both drift and direction of drift
- can be run on wireline, dropped into drill stem and retrieved with an overshot, or dropped into drill stem and recovered when stem is pulled

Glossary

A

adjustable kickoff tool (AKO) *n*: a part of a downhole motor assembly used to kick off or deflect the hole from vertical in a directional hole. The critical equipment is the adjustable kickoff sub (as opposed to a fixed bent sub or bent housing). The kickoff angle can be set in an adjustable sub then pulled and reset without changing tools. It can be used to increase or to decrease hole angle. See *bent housing*.

air drilling *n*: a method of rotary drilling that uses compressed air as the circulating medium. The conventional method of removing cuttings from the wellbore is to use a flow of water or drilling mud. Compressed air removes the cuttings with equal or greater efficiency. The rate of penetration is usually increased considerably when air drilling is used. However, a principal problem in air drilling is the penetration of formation containing water, since the entry of water into the system reduces the ability of the air to remove the cuttings.

American Petroleum Institute (API) *n*: oil trade organization (founded in 1920) that is the leading standardizing organization for oilfield drilling and producing equipment. It maintains departments of transportation, refining, marketing, and production in Washington, DC. It offers publications regarding standards, recommended practices, and bulletins. Address: 1220 L Street NW; Washington, DC 20005; (202) 682-8000.

angle of deviation *n*: also called drift angle and angle of drift. See *deviation*.

angle of drift *n*: also called angle of deviation and drift angle. See *deviation*.

annular space *n*: the space between two concentric circles. In the petroleum industry, it is usually the space surrounding a pipe in the wellbore, or the space between tubing and casing, or the space between tubing and the wellbore; sometimes termed the annulus.

API *abbr*: American Petroleum Institute.

API gravity *n*: the measure of the density or gravity of liquid petroleum products in the United States; derived from relative density in accordance with the following equation:

$$\text{API gravity at } 60°\text{F} = [141.5 \div \text{relative density } 60/60°\text{F}] - 131.5$$

API gravity is expressed in degrees, 10° API being equivalent to 1.0, the specific gravity of water. See *gravity*.

azimuth *n*: 1. in directional drilling, the direction of the wellbore or of the face of a deflection tool in degrees (0° –359°) clockwise from true north. 2. an arc of the horizon measured between a fixed point (such as true north) and the vertical circle passing through the center of an object.

B

baffle plate *n*: 1. a partial restriction, generally a plate, placed to change the direction, guide the flow, or promote mixing within a tank or vessel. 2. a device that is seated on the bit pin, in a tool joint, or in a drill pipe float, used to centralize the lower end of a go-devil while permitting the bypass of drilling fluid. The go-devil contains a surveying instrument.

ball up *v*: 1. to collect a mass of sticky consolidated material, usually drill cuttings, on drill pipe, drill collars, bits, and so forth. A bit with such material attached to it is termed a balled-up bit. Balling up is frequently the result of inadequate pump pressure or insufficient drilling fluid. 2. in reference to an anchor, to fail to hold on a soft bottom, pulling out instead with a large ball of mud attached.

bedding plane *n*: the surface that separates each successive layer of a stratified rock from its preceding layer.

bent housing *n*: a special housing for the positive-displacement downhole mud motor, which is manufactured with a bend of 1° to 3° to facilitate directional drilling.

bent sub *n*: a short, cylindrical device installed in the drill stem between the bottommost drill collar and a downhole motor. Its purpose is to deflect the downhole motor off vertical to drill a directional hole. See *drill stem.*

BHA *abbr*: bottomhole assembly.

bit *n*: the cutting or boring element used in drilling oil and gas wells. The bit consists of a cutting element and a circulating element. The cutting element is steel teeth, tungsten carbide buttons, industrial diamonds, or polycrystalline diamond compacts (PDCs). The circulating element permits the passage of drilling fluid and utilizes the hydraulic force of the fluid stream to improve drilling rates. In rotary drilling, several drill collars are joined to the bottom end of the drill pipe column, and the bit is attached to the end of the drill collars.

bit gauge *n*: a circular ring used to determine whether a bit is of the correct outside diameter. Bit gauges are often used to determine whether the bit has been worn down to a diameter smaller than specifications allow; such a bit is described as undergauge.

bit record *n*: a report that lists each bit used during a drilling operation, giving the type of each, the footage it drilled, and the formation it penetrated.

blowout *n*: an uncontrolled flow of gas, oil, or other well fluids into the atmosphere. A blowout, or gusher, can occur when formation pressure exceeds the pressure applied to it by the column of drilling fluid. A kick warns of an impending blowout. See *formation pressure, gusher, kick.*

bore *n*: 1. the inside diameter of a pipe or a drilled hole. 2. the diameter of the cylinder of an engine. *v*: to penetrate or pierce with a rotary tool.

borehole *n*: the wellbore; the hole made by drilling or boring. See *wellbore.*

bottomhole *n*: the lowest or deepest part of a well. *adj*: pertaining to the bottom of the wellbore.

bottomhole assembly (BHA) *n*: the portion of the drilling assembly below the drill pipe. It can be very simple—composed of only the bit and drill collars—or it can be very complex and made up of several drilling tools.

break out *v*: 1. to unscrew one section of pipe from another section, especially drill pipe while it is being withdrawn from the wellbore. During this operation, the tongs are used to start the unscrewing operation. 2. to separate, as gas from a liquid or water from an emulsion.

build up *v*: to increase the rate of inclination, or drift angle, in the hole in order to bring the hole towards the horizontal. Usually expressed as degrees per 100 ft (30 m).

buoyancy *n*: the apparent loss of weight of an object immersed in a fluid. If the object is floating, the immersed portion displaces a volume of fluid the weight of which is equal to the weight of the object.

bushing *n*: 1. a pipe fitting on which the external thread is larger than the internal thread to allow two pipes of different sizes to be connected together. 2. a removable lining or sleeve inserted or screwed into an opening to limit its size, resist wear or corrosion, or serve as a guide.

C

C *abbr*: Celsius (formerly centigrade). See *Celsius scale*.

casing *n*: steel pipe placed in an oil or gas well as drilling progresses to prevent the wall of the hole from caving in during drilling, to prevent seepage of fluids, and to provide a means of extracting petroleum if the well is productive.

caving *n*: collapsing of the walls of wellbore; also called sloughing.

Celsius scale *n*: the metric scale of temperature measurement used universally by scientists. On this scale, 0 degrees represents the freezing point of water and 100 degrees its boiling point at a barometric pressure of 760 mm. Degrees Celsius are converted to degrees Fahrenheit by using the following equation:

$$°F = \tfrac{9}{5}\,(°C) + 32.$$

The Celsius scale was formerly called the centigrade scale; now, however, the term *Celsius* is preferred in the International System of Units (SI).

cementing *n*: the application of a liquid slurry of cement and water to various points inside or outside the casing.

centigrade scale *n*: see *Celsius scale*.

circulating pressure *n*: the pressure generated by the mud pumps and exerted on the drill stem.

circulation *n*: the movement of drilling fluid out of the mud pits, down the drill stem, up the annulus, and back to the mud pits. See *normal circulation*, *reverse circulation*.

collar *n*: 1. a coupling device used to join two lengths of pipe. A combination collar has left-hand threads in one end and right-hand threads in the other. 2. a drill collar. See *drill collar*.

complete a well *v*: to finish work on a well and bring it to productive status. See *well completion*.

compression *n*: the act or process of squeezing a given volume of gas into a smaller space.

cone *n*: a conical-shaped metal device into which cutting teeth are formed or mounted on a roller cone bit. See *roller cone bit.*

cone bit *n*: a roller bit in which the cutters are conical. See *bit.*

cone offset *n*: the amount by which lines drawn through the center of each cone of a bit fail to meet in the center of the bit. For example, in a roller cone bit with three cones, three lines can be drawn through the center of each cone and extended to the center of the bit. If these cone centerlines do not meet in the bit's center, the cones are said to be offset. In general, bits designed for drilling soft formations have more offset than cones for hard formations, because offset affects the angle at which the bit teeth contact the formation. Since soft formations require a gouging and scraping action by bit teeth, high offset achieves the necessary action.

contract *n*: an agreement, usually written, listing the terms under which services are to be performed. A drilling contract covers such factors as the cost of drilling the well (whether by the foot or by the day), the distribution of expenses between operator and contractor, and the type of equipment to be used.

contract depth *n*: the depth of the wellbore at which a drilling contract is fulfilled.

corrosion *n*: any of a variety of complex chemical or electrochemical processes by which metal is destroyed through reaction with its environment. For example, rust is corrosion. See *corrosion cell.*

corrosion cell *n*: an area on a corrodable substance (usually metal) where corrosion occurs because electrical current is able to flow.

crooked hole *n*: a wellbore that has deviated from the vertical. It usually occurs where there is a section of alternating hard and soft strata steeply inclined from the horizontal.

crooked-hole country *n*: a geographical area in which the subsurface formations are so arranged that it is difficult to drill a hole straight through them. See *crooked hole.*

cutters *n pl*: 1. on a bit used on a rotary rig, the elements on the end (and sometimes the sides) of the bit that scrapes, gouges, or otherwise removes the formation to make hole. 2. the parts of a reamer that actually contact the wall of the hole and open the hole to full gauge. A three-point reamer has three cutters; a six-point reamer has six cutters. Various cutters are available for different formations.

D

daily drilling report *n*: a record made each day of the operations on a working drilling rig and, traditionally, phoned or radioed in to the office of the drilling company every morning. Also called morning report.

deflection *n*: a change in the angle of a wellbore. In directional drilling, it is measured in degrees from the vertical.

depth *n*: 1. the distance to which a well is drilled, stipulated in a drilling contract as contract depth. Total depth is the depth after drilling is finished. 2. on offshore drilling rigs, the distance from the baseline of a rig or a ship to the uppermost continuous deck. 3. the maximum pressure that a diver attains during a dive, expressed in feet (metres) of seawater.

deviation *n*: the inclination of the wellbore from the vertical. The angle of deviation, angle of drift, or drift angle is the angle in degrees that shows the variation from the vertical as revealed by a deviation survey. See *deviation survey*.

deviation survey *n*: an operation made to determine the angle from which a bit has deviated from the vertical during drilling. There are two basic deviation-survey, or drift-survey, instruments: one reveals the angle of deviation only; the other indicates both the angle and the direction of deviation.

diameter *n*: the distance across a circle, measured through its center. In the measurement of pipe diameters, the inside diameter is that of the interior circle and the outside diameter that of the exterior circle.

differential pressure *n*: the difference between two fluid pressures; for example, the difference between the pressure in a reservoir and in a wellbore drilled in the reservoir, or between atmospheric pressure at sea level and at 10,000 feet (3,048 metres). Also called pressure differential.

differential sticking *n*: a condition in which the drill stem becomes stuck against the wall of the wellbore because part of the drill stem (usually the drill collars) has become embedded in the filter cake. Necessary conditions for differential-pressure sticking, or wall sticking, are a permeable formation and a pressure differential across a nearly impermeable filter cake and drill stem. Also called wall sticking. See *differential pressure, filter cake.*

dip *n*: also called formation dip. *See formation dip.*

directional drilling *n*: intentional deviation of a wellbore from the vertical. Although wellbores are normally drilled vertically, it is sometimes necessary or advantageous to drill at an angle from the vertical. Controlled directional drilling makes it possible to reach subsurface areas laterally remote from the point where the bit enters the earth. It often involves the use of deflection tools.

dogleg *n*: 1. an abrupt change of direction in the wellbore, frequently resulting in the formation of a key seat. See *key seat.* 2. a sharp bend permanently put in an object such as a pipe.

dolomite *n*: a type of sedimentary rock similar to limestone but rich in magnesium carbonate; sometimes a reservoir rock for petroleum.

downhole *adj, adv*: pertaining to the wellbore.

downhole motor *n*: a drilling tool made up in the drill string directly above the bit. It causes the bit to turn while the drill string remains fixed. It is used most often as a deflection tool in directional drilling, where it is made up between the bit and a bent sub (or, sometimes, the housing of the motor itself is bent). Two principal types of downhole motor are the positive-displacement motor and the downhole turbine motor. Also called mud motor.

drift *n*: the deviation of a wellbore from vertical; usually measured in degrees. For example, if a wellpipe deviates from vertical by three degrees, it has a three-degree drift. *v*: 1. to move slowly out of alignment, off center, or out of register. 2. to gauge or measure pipe by means of a mandrel passed through it to ensure the passage of tools, pumps, etc.

drift angle *n*: also called angle of deviation and angle of drift. See *deviation.*

drift diameter *n*: in drilling, the effective hole size.

drift indicator *n*: a device dropped or run down the drill stem on a wireline to a point just above the bit to measure the inclination of the well off vertical at that point. It does not measure the direction of the inclination.

drift survey *n*: see *deviation survey.*

drill *v*: to bore a hole in the earth, usually to find and remove subsurface formation fluids such as oil and gas.

drill bit *n*: the cutting or boring element used for drilling. See *bit.*

drill collar *n*: a heavy, thick-walled tube, usually steel, used between the drill pipe and the bit in the drill stem to provide a pendulum effect to the drill stem and weight on the bit.

driller *n*: the employee directly in charge of a drilling or workover rig and crew. The driller's main duty is operation of the drilling and hoisting equipment, but this person is also responsible for downhole condition of the well, operation of downhole tools, and pipe measurements.

driller's log *n*: a record that describes each formation encountered and lists the drilling time relative to depth, usually in 5- to 10-foot (1.5- to 3-metre) intervals.

drill floor *n*: also called rig floor or derrick floor. See *rig floor.*

drilling contract *n*: an agreement made between a drilling company and an operating company to drill and complete a well, setting forth the obligation of each party, compensation, identification, method of drilling, depth to be drilled, and so on.

drilling contractor *n*: an individual or group of individuals who own a drilling rig and contract their services for drilling wells.

drilling engineer *n*: an engineer who specializes in the technical aspects of drilling.

drilling fluid *n*: circulating fluid, one function of which is to lift cuttings out of the wellbore and to the surface. It also serves to cool the bit and to counteract downhole formation pressure. Although a mixture of barite, clay, water, and other chemical additives is the most common drilling fluid, wells can also be drilled by using air, gas, water, or oil-base mud as the drilling mud. Also called circulating fluid, drilling mud. See *mud.*

drilling mud *n*: a specially compounded liquid circulated through the wellbore during rotary drilling operations. See *drilling fluid, mud.*

drilling rate *n*: the speed with which the bit drills the formation; usually called the rate of penetration (ROP).

drilling rig *n*: see *rig.*

drill pipe *n*: heavy seamless tubing used to rotate the bit and circulate the drilling fluid. Joints of pipe approximately 30 feet (9 metres) long are coupled together by means of tool joints.

drill pipe float *n*: a check valve installed in the drill stem that allows mud to be pumped down the drill stem but prevents flow back up the drill stem.

drill stem *n*: all members in the assembly used for rotary drilling from the swivel to the bit, including the kelly, drill pipe and tool joints, drill collars, stabilizers, and various specialty items. Compare *drill string*.

drill string *n*: the column, or string, of drill pipe with attached tool joints that transmits fluid and rotational power from the kelly to the drill collars and bit. Often, especially in the oil patch, the term is loosely applied to both drill pipe and drill collars. Compare *drill stem*.

drop off *v*: to reduce the rate of inclination, or drift angle, in the hole. The objective being to bring the hole back towards vertical. Usually expressed as degrees per 100 ft (30 m).

E

electric well log *n*: a record of certain electrical characteristics (such as resistivity and conductivity) of formations traversed by the borehole. It is made to identify the formations, determine the nature and amount of fluids they contain, and estimate their depth. Also called an electric log or electric survey.

F

F *abbr*: Fahrenheit. See *Fahrenheit scale*.

Fahrenheit scale *n*: a temperature scale devised by Gabriel Fahrenheit, in which 32° represents the freezing point and 212° the boiling point of water at standard sea-level pressure. Fahrenheit degrees may be converted to Celsius degrees by using the following formula:

$$°C = 5/9\,(°F - 32)$$

fatigue *n*: the tendency of a material such as metal to break under repeated cyclic loading at a stress considerably less than the tensile strength shown in a static test.

filter cake *n*: 1. compacted solid or semisolid material remaining on a filter after pressure filtration of mud with a standard filter press. Thickness of the cake is reported in thirty-seconds of an inch or in millimetres. 2. the layer of concentrated solids from the drilling mud or cement slurry that forms on the walls of the borehole opposite permeable formations; also called wall cake or mud cake.

fish *n*: an object that is left in the wellbore during drilling or workover operations and that must be recovered before work can proceed. It can be anything from a piece of scrap metal to a part of the drill stem. *v*: 1. to recover from a well any equipment left there during drilling operations, such as a lost bit or drill collar or part of the drill string. 2. to remove from an older well certain pieces of equipment (such as packers, liners, or screen liner) to allow reconditioning of the well.

fluid loss *n*: the unwanted migration of the liquid part of the drilling mud or cement slurry into a formation, often minimized or prevented by the blending of additives with the mud or cement.

formation *n*: a bed or deposit composed throughout of substantially the same kind of rock; often a lithologic unit. Each formation is given a name, frequently as a result of the study of the formation outcrop at the surface and sometimes based on fossils found in the formation.

formation dip *n*: the angle at which a formation bed inclines away from the horizontal.

formation fluid *n*: fluid (such as gas, oil, or water) that exists in a subsurface rock formation.

formation pressure *n*: the force exerted by fluids in a formation, recorded in the hole at the level of the formation with the well shut in. Also called reservoir pressure or shut-in bottomhole pressure.

friction *n*: resistance to movement created when two surfaces are in contact. When friction is present, movement between the surfaces produces heat.

full-gauge bit *n*: a bit that has maintained its original diameter.

full-gauge hole *n*: a wellbore drilled with a full-gauge bit. Also called a true-to-gauge hole.

G

gauge *n*: 1. the diameter of a bit or the hole drilled by the bit. 2. a device (such as a pressure gauge) used to measure some physical property. *v*: to measure size, volume, or other measurable property.

geologist *n*: a scientist who gathers and interprets data pertaining to the strata of the earth's crust.

Geolograph™ *n*: trade name for a patented device that automatically records the rate of penetration and depth during drilling.

go-devil *n*: 1. a device that is inserted into a pipeline for the purpose of cleaning; a line scraper. Also called a pig. 2. a device that is lowered into the borehole of a well for various purposes such as enclosing surveying instruments, detonating instruments, and the like. *v*: to drop or pump a device down the borehole, usually through drill pipe or tubing.

gravity *n*: the attraction exerted by the earth's mass on objects at its surface; the weight of a body. See *API gravity*, *specific gravity*.

gusher *n*: an oilwell that has come in with such great pressure that the oil jets out of the well like a geyser. In reality, a gusher is a blowout and is extremely wasteful of reservoir fluids and drive energy. In the early days of the oil industry, gushers were common and many times were the only indication that a large reservoir of oil and gas had been struck. See *blowout*.

H

hammer drill *n*: a drilling tool that, when placed in the drill stem just above a roller cone bit, delivers high-frequency percussion blows to the rotating bit. Hammer drilling combines the basic features of rotary and cable-tool drilling (i.e, bit rotation and percussion).

hardfacing *n*: an extremely hard material, usually crushed tungsten carbide, that is applied to the outside surfaces of tool joints, drill collars, stabilizers, and other rotary drilling tools to minimize wear when they are in contact with the wall of the hole.

hole *n*: 1. in drilling operations, the wellbore or borehole. See *borehole*, *wellbore*. 2. an opening that is made purposely or accidentally in any solid substance.

horizontal drilling *n*: deviation of the borehole at least 80° from vertical so that the borehole penetrates a productive formation in a manner parallel to the formation. A single horizontal hole can effectively drain a reservoir and eliminate the need for several vertical boreholes.

hydrostatic pressure *n*: the force exerted by a body of fluid at rest. It increases directly with the density and the depth of the fluid and is expressed in pounds per square inch or kilopascals. The hydrostatic pressure of fresh water is 0.433 pounds per square inch per foot (9.792 kilopascals per metre) of depth. In drilling, the term refers to the pressure exerted by the drilling fluid in the wellbore. In a water drive field, the term refers to the pressure that may furnish the primary energy for production.

I

ID *abbr*: inside diameter.

in. *abbr*: inch.

inertia *n*: the tendency of an object having mass to resist a change in velocity.

inside diameter (ID) *n*: distance across the interior of a circle, especially in the measurement of pipe. See *diameter*.

J

junk *n*: metal debris lost in a hole. Junk may be a lost bit, pieces of a bit, milled pieces of pipe, wrenches, or any relatively small object that impedes drilling and must be fished out of the hole. *v*: to abandon (as a nonproductive well).

K

kelly *n*: the heavy steel member, four- or six-sided, suspended from the swivel through the rotary table and connected to the top joint of drill pipe to turn the drill stem as the rotary table turns. It has a bored passageway that permits fluid to be circulated into the drill stem and up the annulus, or vice versa.

keyseat *n*: 1. an undergauge channel or groove cut in the side of the borehole and parallel to the axis of the hole. A keyseat results from the rotation of pipe on a sharp bend in the hole. 2. a groove cut parallel to the axis in a shaft or a pulley bore.

kick *n*: an entry of water, gas, oil, or other formation fluid into the wellbore during drilling. It occurs because the pressure exerted by the column of drilling fluid is not great enough to overcome the pressure exerted by the fluids in the formation drilled. If prompt action is not taken to control the kick, or kill the well, a blowout may occur.

L

lb *abbr:* pound.

lb/ft^3 *abbr*: pounds per cubic foot.

lease *n*: 1. a legal document executed between a landowner, as lessor, and a company or individual, as lessee, that grants the right to exploit the premises for minerals or other products. 2. the area where production wells, stock tanks, separators, LACT units, and other production equipment are located.

location *n*: the place where a well is drilled; also called well site.

log *n*: a systematic recording of data, such as a driller's log, mud log, electrical well log, or radioactivity log. Many different logs are run in wells to obtain various characteristics of downhole formations. *v*: to record data.

logging devices *n pl*: any of several electrical acoustical, mechanical, or radioactivity devices that are used to measure and record certain characteristics or events that occur in a well that has been or is being drilled.

logging while drilling (LWD) *n*: logging measurements obtained by measurement-while-drilling techniques as the well is being drilled.

LWD *abbr*: logging while drilling.

M

make a trip *v*: to hoist the drill stem out of the wellbore to perform one of a number of operations, such as changing bits, taking a core, and so forth, and then to return the drill stem to the wellbore.

make hole *v*: to deepen the hole made by the bit; to drill ahead.

make up *v*: 1. to assemble and join parts to form a complete unit (as to make up a string of casing). 2. to screw together two threaded pieces. 3. to mix or prepare (as to make up a tank of mud). 4. to compensate for (as to make up for lost time.)

mandrel *n*: a cylindrical bar, spindle, or shaft around which other parts are arranged or attached or that fits inside a cylinder or tube.

measurement while drilling (MWD) *n*: 1. directional and other surveying during routine drilling operations to determine the angle and direction by which the wellbore deviates from the vertical. 2. any system of measuring downhole conditions during routine drilling operations.

metre (m) *n*: the fundamental unit of length in the international system of measurement (SI). It is equal to about 3.28 feet, 39.37 inches, or 100 centimetres.

moment *n*: a turning effect created by a force, F, acting at a perpendicular distance, S, from the center of rotation; the product of a force and a distance to a particular axis or point.

mud *n*: the liquid circulated through the wellbore during rotary drilling and workover operations. In addition to its function of bringing cuttings to the surface, drilling mud cools and lubricates the bit and the drill stem, protects against blowouts by holding back subsurface pressures, and deposits a mud cake on the wall of the borehole to prevent loss of fluids to the formation. Although it was originally a suspension of earth solids (especially clays) in water, the mud used in modern drilling operations is a more complex three-phase mixture of liquids, reactive solids, and nonreactive solids. The liquid phase may be fresh water, diesel oil, or crude oil and may contain one or more conditioners. See *drilling fluid, drilling mud.*

mud column *n*: the borehole when it is filled or partially filled with drilling mud.

mud weight *n*: a measure of the density of a drilling fluid expressed as pounds per gallon, pounds per cubic foot, or kilograms per cubic metre. Mud weight is directly related to the amount of pressure the column of drilling mud exerts at the bottom of the hole.

MWD *abbr*: measurement while drilling.

N

nomograph *n*: a chart that presents an equation containing a number of variables in the form of scales so that a straight line cuts the scales at values of the variables satisfying the equation.

normal circulation *n*: the smooth, uninterrupted circulation of drilling fluid down the drill stem, out the bit, up the annular space between the pipe and the hole, and back to the surface. Compare *reverse circulation.*

O

OD *abbr*: outside diameter.

operator *n*: the person or company, either proprietor or lessee, actually operating an oilwell or lease. Generally, the oil company by whom the drilling contractor is engaged. Compare *unit operator.*

optimization *n*: the manner of planning and drilling a well so that the most usable hole will be drilled for the least money.

outside diameter (OD) *n*: the distance across the exterior circle, especially in the measurement of pipe. See *diameter.*

overshot *n*: a fishing tool that is attached to tubing or drill pipe and lowered over the outside wall of pipe or sucker rods lost or stuck in the wellbore. A friction device in the overshot, usually either a basket or a spiral grapple, firmly grips the pipe, allowing the lost fish to be pulled from the hole.

P

packed-hole assembly *n*: a drill stem that consists of stabilizers and special drill collars and is used to maintain the proper angle and course of the hole. This assembly is often necessary in crooked-hole country. See *crooked-hole country.*

packed pendulum assembly *n*: a bottomhole assembly in which pendulum-length collars are swung below a regular packed-hole assembly. The pendulum portion of the assembly is used to reduce hole angle. It is then removed, and the packed-hole assembly is run above the bit. See *packed-hole assembly, pendulum assembly.*

pay sand *n*: the producing formation, often one that is not even sandstone. It is also called pay, pay zone, and producing zone.

pay zone *n*: see *pay sand.*

pcf *abbr*: pounds per cubic foot.

PDC *abbr*: polycrystalline diamond compact.

PDC bit *n*: a special type of diamond drilling bit that does not use roller cones. Instead, polycrystalline diamond inserts are embedded into a matrix on the bit. PDC bits are often used to drill very hard, abrasive formations, but also find use in drilling medium and soft formations.

pendulum assembly *n*: a bottomhole assembly composed of a bit and several large-diameter drill collars; it may have one or more stabilizers installed in the drill collar string. The assembly works on the principle of the pendulum effect. See *pendulum effect.*

pendulum effect *n*: the tendency of the drill stem—bit, drill collars, drill pipe, and kelly—to hang in a vertical position due to the force of gravity.

penetration rate *n*: see *rate of penetration.*

pickup position *n*: the point in drilling at which the floor crew can latch the elevators around the pipe for coming out of the hole.

pipe *n*: a long, hollow cylinder, usually steel, through which fluids are conducted. Oilfield tubular goods are casing (including liners), drill pipe, tubing, or line pipe. Casing, tubing, and drill pipe are designated by external diameter. Because lengths of pipe are joined by external-diameter couplings threaded by standard tools, an increase in the wall thickness can be obtained only by decreasing the internal diameter. Thus, the external diameter is the same for all weights of the same-size pipe. Weight is expressed in pounds per foot or kilograms per metre. Grading depends on the yield strength of the steel.

pounds per gallon (ppg) *n*: a measure of the density of a fluid (such as drilling mud).

producing zone *n*: the zone or formation from which oil or gas is produced. See *pay sand.*

production *n*: 1. the phase of the petroleum industry that deals with bringing the well fluids to the surface and separating them and with storing, gauging, and otherwise preparing the product for the pipeline. 2. the amount of oil or gas produced in a given period.

R

rate of penetration (ROP) *n*: a measure of the speed at which the bit drills into formations, usually expressed in feet (metres) per hour or minutes per foot (metre).

rathole *n*: 1. a hole in the rig floor, some 30 to 40 feet (9 to 12 metres) deep, which is lined with casing that projects above the floor, into which the kelly and the swivel are placed when hoisting operations are in progress. 2. a hole of a diameter smaller than the main hole and drilled in the bottom of the main hole. *v*: to reduce the size of the wellbore and drill ahead.

ream *v*: to enlarge the wellbore by drilling it again with a special bit. Often a rathole is reamed or opened to the same size as the main wellbore. See *rathole.*

reamer *n*: a tool used in drilling to smooth the wall of a well, enlarge the hole to the specified size, help stabilize the bit, straighten the wellbore if kinks or doglegs are encountered, and drill directionally. See *ream.*

relative density *n*: 1. the ratio of the weight of a given volume of a substance at a given temperature to the weight of an equal volume of a standard substance at the same temperature. For example, if 1 cubic inch of water at 39°F (3.9°C) weighs 1 unit and 1 cubic inch of another solid or liquid at 39°F weighs 0.95 unit, then the relative density of the substance is 0.95. In determining the relative density of gases, the comparison is made with the standard of air or hydrogen. 2. the ratio of the mass of a given volume of a substance to the mass of a like volume of a standard substance, such as water or air.

reverse circulation *n*: the course of drilling fluid downward through the annulus and upward through the drill stem, in contrast to normal circulation in which the course is downward through the drill stem and upward through the annulus. Seldom used in open hole, but frequently used in workover operations. Also referred to as "circulating the short way," since returns from bottom can be obtained more quickly than in normal circulation. Compare *normal circulation.*

rig *n*: the derrick or mast, drawworks, and attendant surface equipment of a drilling or workover unit.

rig floor *n*: the area immediately around the rotary table and extending to each corner of the derrick or mast—the area immediately above the substructure on which the drawworks, the rotary table, and so forth rest. Also called derrick floor, drill floor.

roller cone bit *n*: a drilling bit made of two, three, or four cones, or cutters, that are mounted on extremely rugged bearings. Also called rock bits. The surface of each cone is made up of rows of steel teeth or rows of tungsten carbide inserts.

ROP *abbr*: rate of penetration.

rotary drilling *n*: a drilling method in which a hole is drilled by a rotating bit to which a downward force is applied. The bit is fastened to and rotated by the drill stem, which also provides a passageway through which the drilling fluid is circulated. Additional joints of drill pipe are added as drilling progresses.

rotary speed *n*: the speed, measured in revolutions per minute, at which the rotary table is operated.

rotor *n*: 1. a device with vanelike blades attached to a shaft. The device turns or rotates when the vanes are struck by a fluid directed there by a stator. 2. the rotating part of an induction-type alternating current electric motor.

rpm *abbr*: revolutions per minute.

run casing *v*: to lower a string of casing into the hole. Also called to run pipe.

run in *v*: to go into the hole with tubing, drill pipe, and so forth.

S

sand line *n*: a small diameter, flexible steel line kept on a spool mounted on the drawworks. It is used to lower and recover survey instruments down the borehole or casing.

sedimentary rock *n*: a rock composed of materials that were transported to their present position by wind or water. Sandstone, shale, and limestone are sedimentary rocks.

set casing *v*: to run and cement casing at a certain depth in the wellbore. Sometimes, the term set pipe is used in reference to setting casing.

skid the rig *v*: to move a rig with a standard derrick from the location of a lost or completed hole preparatory to starting a new hole. Skidding the rig allows the move to be accomplished with little or no dismantling of equipment.

sleeve *n*: a tubular part designed to fit over another part.

sloughing *n*: (pronounced "sluffing"). Also called caving. See *caving*.

specific gravity *n*: see *relative density*.

square drill collar *n*: a special drill collar, square but with rounded edges, used to control the straightness or direction of the hole; often part of a packed-hole assembly.

stabilizer *n*: 1. a tool placed near the bit, and often just above it in the drilling assembly and used to change the deviation angle in a well by controlling the location of the contact point between the hole and the drill collars. Conversely, stabilizers are used to maintain correct hole angle. See *packed-hole assembly*. 2. a vessel in which hydrocarbon vapors are separated from liquids.

stand *n*: the connected joints of pipe racked in the derrick or mast during a trip. The usual stand is about 90 feet long (about 27 metres), which is three lengths of drill pipe screwed together (a thribble).

straight hole *n*: a hole that is drilled vertically. The total hole angle is restricted, and the hole does not change direction rapidly—no more than 3° per 100 feet (30.48 metres) of hole.

stress *n*: a force that, when applied to an object, distorts or deforms it.

string *n*: the entire length of casing, tubing, sucker rods, or drill pipe run into a hole.

stuck pipe *n*: drill pipe, drill collars, casing, or tubing that has inadvertently become immovable in the hole. It may occur when drilling is in progress, when casing is being run in the hole, or when the drill pipe is being hoisted.

sucker rod *n*: a special steel pumping rod. Several rods screwed together make up the mechanical link from the beam pumping unit on the surface to the sucker rod pump at the bottom of a well. Sucker rods are threaded on each end and manufactured to dimension standards and metal specifications set by the petroleum industry. Lengths are from 25 or 30 feet (7.6 or 9.1 metres), diameter varies from ½ to 1⅛ inches (12 to 30 millimetres). There is also a continuous sucker rod (trade name: Corod™).

T

tapered string *n*: drill pipe, tubing, sucker rods, and so forth with a diameter near the top of the well larger than the diameter below.

TD abbr: total depth, usually the final drilled depth.

tensile *adj*: of or relating to tension.

tensile strength *n*: the greatest longitudinal stress that a metal can bear without tearing apart. Tensile strength of a metal is greater than yield strength.

tensile stress *n*: stress developed by a material bearing a tensile load. See *stress*.

tension *n*: the condition of a string, wire, pipe, or rod that is stretched between two points.

tongs *n pl*: the large wrenches used for turning in making up or breaking out drill pipe, casing, tubing, or other pipe; variously called casing tongs, pipe tongs, and so forth, according to the specific use. Power tongs are pneumatically or hydraulically operated tools that serve to spin the pipe up tight and, in some instance, to apply the final makeup torque.

tool joint *n*: a heavy coupling element for drill pipe made of special alloy steel. Tool joints have coarse, tapered threads and seating shoulders designed to sustain the weight of the drill stem, withstand the strain of frequent coupling and uncoupling, and provide a leakproof seal. The male section of the joint, or the pin, is attached to one end of a length of drill pipe, and the female section, or the box, is attached to the other end. The tool joint may be welded to the end of the pipe, screwed on, or both. A hard-metal facing is often applied in a band around the outside of the tool joint to enable it to resist abrasion from the walls of the borehole.

torque *n*: the turning force that is applied to a shaft or other rotary mechanism to cause it to rotate or tend to do so. Torque is measured in foot-pounds, joules, newton-metres, and so forth.

torsion *n*: twisting deformation of a solid body about an axis in which lines that were initially parallel to the axis become helices. Torsion is produced when part of the pipe turns or twists in one direction while the other part remains stationary or twists in the other direction.

total depth (TD) *n*: the maximum depth reached in a well.

tour (pronounced "tower") *n*: a working shift for drilling crew or other oilfield workers. Some tours are 8 hours; the three daily tours are called daylight, evening, and graveyard (or morning). Often 12-hour tours are used, especially on offshore rigs; they are called simply day tour and night tour.

transition zone *n*: 1. the area in which underground pressures begin to change from normal to abnormally high as a well is being deepened. 2. the areas in the drill stem near the point where drill pipe is made up on drill collars.

trip *n*: the operation of hoisting the drill stem from and returning it to the wellbore. *v*: shortened form of "make a trip." See *make a trip*.

tripping *n*: the operation of hoisting the drill stem out of and returning it to the wellbore. See *make a trip*.

true vertical depth (TVD) *n*: the depth of a well measured from the surface straight down to the bottom of the well. The true vertical depth of a well may be quite different from its actual measured depth, because wells are very seldom drilled exactly vertical.

tubing *n*: small-diameter pipe that is run into a well to serve as a conduit for the passage of oil and gas to the surface.

tungsten carbide *n*: a fine, very hard, gray crystalline powder, a compound of tungsten and carbon. This compound is bonded with cobalt or nickel in cemented carbide compositions and used for cutting tools, abrasives, and dies.

turn *n*. a change in bearing (azimuth) of the hole. Usually spoken of as a right or left turn from the established hole drift.

TVD *abbr*: true vertical depth.

twistoff *n*: a complete break in pipe caused by metal fatigue.

U

undergauge bit *n*: a bit whose outside diameter is worn to the point at which it is smaller than it was when new. A hole drilled with an undergauge bit is said to be undergauge.

unit operator *n*: the oil company in charge of development and production in an oilfield in which several companies have joined to produce the field. Also called crew chief, rig operator.

V

vertical *n*: an imaginary line at right angles to the plane of the horizon. *adj*: of a wellbore, straight, not deviated.

vibration dampener *n*: 1. a device, positioned in the drill stem between the bit and the drill collars, that absorbs impact loads and vibration from the up-and-down motion of the drill stem. Vibration dampeners are designed to transmit torque while absorbing reciprocative loads that decrease the efficiency of the drill bit. Also called a Shock Sub. 2. a device affixed to an engine crankshaft to minimize stresses that result from torsional vibration of the crankshaft.

W

wall cake *n*: also called filter cake or mud cake. See *filter cake.*

wall sticking *n*: see *differential sticking.*

washover *n*: the operation during which stuck drill stem or tubing is freed using washover pipe.

wash over *v*: to release pipe that is stuck in the hole by running washover pipe. The washover pipe must have an outside diameter small enough to fit into the borehole but an inside diameter large enough to fit over the outside diameter of the stuck pipe. A rotary shoe, which cuts away the formation, mud, or whatever is sticking the pipe, is made up on the bottom joint of the washover pipe, and the assembly is lowered into the hole. Rotation of the assembly frees the stuck pipe. Several washovers may have to be made if the stuck portion is very long.

washover assembly *n*: see *washover pipe.*

washover pipe *n*: an accessory used in fishing operations to go over the outside of tubing or drill pipe stuck in the hole because of cuttings, mud, and so forth that have collected in the annulus. The washover pipe cleans the annular space and permits recovery of the pipe. It is sometimes called washpipe.

washpipe *n*: 1. a short length of surface-hardened pipe that fits inside the swivel and serves as a conduit for drilling fluid through the swivel.

weight on bit (WOB) *n*: the difference between the net weight of the entire drill stem and the reduced weight resulting when the bit is resting on bottom.

well *n*: 1. the hole made by the drilling bit, which can be open, cased, or both. Also called borehole, hole, or wellbore. 2. productive oil, gas, or water well. Strictly speaking, a dry hole or an abandoned borehole, whether open or cased, should not be called a well.

wellbore *n*: a borehole; the hole drilled by the hit. A wellbore may have casing in it or it may be open (uncased), or a portion of it may be cased, and a portion of it may be open. Also called borehole or hole.

well completion *n*: the activities and methods necessary to prepare a well for the production of oil and gas, the method by which a flow line for hydrocarbons is established between the reservoir and the surface. The method of well completion used by the operator depends on the individual characteristics of the producing formation or formations. Such techniques include open-hole completions, sand-exclusion completions, tubingless completions, multiple completions, and miniaturized completions.

well logging *n*: the recording of information about subsurface geologic formations. Logging includes records kept by the driller and records of mud and cutting analyses, core analysis, drill stem tests, and electric, acoustic, and radioactivity procedures.

well site *n*: see *location.*

wildcat *n*: 1. a well drilled in an area where no oil or gas production exists. 2. (nautical) the geared sheave of a windlass used to pull anchor chain. *v*: to drill wildcat wells.

wireline *n*: a slender, rodlike or threadlike piece of metal usually small in diameter, that is used for lowering special tools (such as logging sondes, perforating guns, and so forth) into the well. Compare *wire rope.*

wireline operations *n pl*: activities in drilling and production that involve the use of devices put into the hole by wireline.

wire rope *n*: a cable composed of steel wires twisted around a central core of fiber or steel wire to create a rope of great strength and considerable flexibility. Wire rope is used as drilling line (in rotary and cable-tool rigs), coring line, servicing line, winch line, and so on. It is often called cable or wireline; however, wireline is a single, slender metal rod, usually very flexible. Compare *wireline.*

WOB *abbr*: weight on bit.

Z

zone *n*: a rock stratum that is different from or distinguished from another stratum (e.g., a pay zone).

Review Questions

LESSONS IN ROTARY DRILLING

Unit II, Lesson 3: Drilling a Straight Hole

True or False

Put a T for *true* or an F for *false* in the blank next to each statement.

________ 1. If a driller holds the kelly vertical when starting the hole, the hole will be drilled straight to total depth.

________ 2. The first gyroscopic directional survey instrument was not developed until the 1960s.

________ 3. Applying more weight on bit to increase the rate of penetration can cause an increase in hole deviation.

________ 4. A dogleg is a rapid change in the direction of the wellbore.

________ 5. Usually, doglegs do not cause problems and can therefore be ignored.

________ 6. Almost always, a hole should never deviate more than 3 degrees from vertical.

________ 7. A change of 3 degrees of deviation occurring in a 100-ft (30-m) section of hole can make the hole unusable.

________ 8. Modern drilling contracts strictly limit hole deviation to a maximum of 3 degrees from vertical.

Matching

Write the letter of the correct definition in the blank next to each term.

Terms

________ 9. straight hole

________ 10. usable borehole

________ 11. dogleg

________ 12. keyseat

Definitions

a. a rapid change in the direction of the borehole

b. a groove cut into a dogleg by the drill string

c. a full-gauge hole, free of doglegs, keyseats, ledges, offsets, and spirals

d. a hole that stays within the boundary of a cone designated by the operator and does not change direction rapidly

Multiple Choice

Pick the *best* answer from the choices and place the letter of that answer in the blank provided.

________ 13. Problems caused by severe doglegs and keyseats include—
a. drill pipe fatigue.
b. stuck drill string and casing.
c. bad cement jobs.
d. all of the above

________ 14. Formation dip is—
a. an anticline.
b. an upward slope.
c. a downward slope.
d. the inclination of formation bedding planes.

________ 15. In formations with dips of 40 degrees or less, the bit tends to drill—
a. downdip.
b. updip.
c. straight through the dip.
d. none of the above

________ 16. In formations with dips of 40 degrees or more, the bit tends to drill—
a. downdip.
b. updip.
c. straight through the dip.
d. none of the above

________ 17. Which of the following is *not* generally used as a bottomhole assembly (BHA) to control deviation?
a. a drill string with no drill collars
b. a packed-hole assembly
c. a pendulum assembly
d. a packed pendulum assembly

________ 18. The pendulum effect is—
a. the tendency of the bit to drill updip.
b. the bit's tendency to drill a spiral hole.
c. the tendency of the drill string to hang in a vertical position.
d. the tendency of the drill string to hang in a horizontal position.

________ 19. A packed-hole assembly (PHA) consists of—
a. slick drill collars, i.e., a collar string with no stabilizers.
b. one stabilizer placed at least three stands above the bit.
c. drill collars and stabilizers that are no more than ⅛ in. (3 mm) smaller than borehole diameter.
d. one large stabilizer placed just above the bit.

________ 20. A packed pendulum assembly is—
   a. placing from one to three drill collars below the regular PHA.
   b. placing from one to three drill pipe joints below the regular PHA.
   c. placing a large stabilizer just above the bit.
   d. placing a small stabilizer just above the bit.

Fill in the Blanks

Fill in the blanks with the appropriate word or phrase. Pick the correct term from the list below.

square bar	retard
less	outside diameter
differential sticking	inside diameter
wall thickness	transition zone
pressure drop	weight

Drill collars supply (21.) ______ to the bit for drilling. Their weight varies considerably and depends on the (22.) ________ ________ and (23.) ________ ________ of the drill collar. Drill collars with small IDs create more (24.) __________ _____ than collars with large IDs. Operators should select drill collars with the largest OD and maximum permissible (25.) ____ _________ that can be safely run in the hole and fished out in case of trouble. A (26.) _____________ ________ from drill collars to drill pipe is assembled by: reducing drill collar sizes in the upper end of the drill collar string, using heavy-walled or heavyweight drill pipe at the bottom of the drill pipe string, or by combining the two previous assembly methods. Drill collars weigh (27.) _______ in drilling mud than in air. A square drill collar is a (28.) ____________ _______ with rounded corners. A square drill collar can (29.) _______ the change of angle deviation in a well. Spiral drill collars can be used to combat (30.) ____________ ___________.

True or False

Put a T for *true* or an F for *false* in the blank next to each statement.

________ 31. Stabilizers are an integral part of a bottomhole assembly (BHA).

________ 32. Soft formations usually require stabilizers with small wall contact areas.

________ 33. All but one of the following are types of stabilizers: rotating blade, nonrotating sleeve, and rolling-cutter reamer.

________ 34. Integral blade stabilizers are usually employed in soft and very soft formations.

________ 35. A nonrotating sleeve stabilizer is most effective in hard formations.

________ 36. Operators should run at least two stabilizers directly on top of a reamer in severe crooked-hole conditions.

Multiple Choice

Pick the *best* answer from the choices and place the letter of that answer in the blank provided.

_________ 37. A vibration dampener—
a. minimizes bounce and vibration as the bit drills.
b. helps maintain constant weight on bit.
c. reduces stress on the drill string.
d. all of the above

_________ 38. In measurement while drilling (MWD) instruments, data is transmitted by downhole detectors to the surface by—
a. mud pulse telemetry.
b. wireline.
c. both a and b
d. none of the above

_________ 39. Deviation recorders measure—
a. inclination from vertical and the direction of the inclination.
b. direction of inclination only.
c. inclination from vertical only.
d. none of the above

_________ 40. To prevent the drill string from sticking while running a survey instrument, the driller should—
a. rotate the drill string slowly and raise and lower it.
b. raise but not lower the drill string.
c. make sure the drill string is stationary during the whole process.
d. keep the drill string moving during the time the survey is taken.

Answers to Review Questions

LESSONS IN ROTARY DRILLING

Unit II, Lesson 3: Drilling a Straight Hole

True or False

1. F
2. F
3. T
4. T
5. F
6. F
7. T
8. F

Matching

9. d
10. c
11. a
12. b

Multiple Choice

13. d
14. d
15. b
16. a
17. a
18. c
19. c
20. a

Fill in the Blanks

21. weight
22. outside diameter
23. inside diameter
24. pressure drop
25. wall thickness
26. transition zone
27. less
28. square bar
29. retard
30. differential sticking

True or False

31. T
32. F
33. F
34. F
35. T
36. T

Multiple Choice

37. d
38. c
39. c
40. a